Atmospheric Science: Methods, Models and Applications

Atmospheric Science: Methods, Models and Applications

Edited by **Smith Paul**

New York

Published by Callisto Reference,
106 Park Avenue, Suite 200,
New York, NY 10016, USA
www.callistoreference.com

Atmospheric Science: Methods, Models and Applications
Edited by Smith Paul

International Standard Book Number:978-1-63239-626-6 (Hardback)

The publisher's policy is to use permanent paper from mills that operate a sustainable forestry policy. Furthermore, the publisher ensures that the text paper and cover boards used have met acceptable environmental accreditation standards.

Trademark Notice: Registered trademark of products or corporate names are used only for explanation and identification without intent to infringe.

Printed in the United States of America.

Contents

Preface

The world is advancing at a fast pace like never before. Therefore, the need is to keep up with the latest developments. This book was an idea that came to fruition when the specialists in the area realized the need to coordinate together and document essential themes in the subject. That's when I was requested to be the editor. Editing this book has been an honour as it brings together diverse authors researching on different streams of the field. The book collates essential materials contributed by veterans in the area which can be utilized by students and researchers alike.

Different layers of gases in combination constitute atmosphere, these layers become thinner with the increase in height. Atmospheric science is a wider term used for studying earth's atmosphere, chemical reactions in stratosphere, gaseous exchange with the biosphere, lifecycle of aerosols, regional and global air pollution, and fluctuations in atmospheric composition. This book discusses some important topics such as air pollution, gas dispersion in atmosphere, seasonal variations, atmospheric chemistry, atmospheric physics, air quality, climate interactions, meteorology, etc. This book includes some of the vital pieces of work being conducted across the world, on various topics related to this field. It will provide comprehensive knowledge to the readers. The various studies that are constantly contributing towards advancing technologies and evolution of this field are examined in detail. It is a beneficial read for climatologists, meteorologists, planetologists, professionals and students.

Each chapter is a sole-standing publication that reflects each author's interpretation. Thus, the book displays a multi-facetted picture of our current understanding of applications and diverse aspects of the field. I would like to thank the contributors of this book and my family for their endless support.

Editor

Meteorological Modeling Using the WRF-ARW Model for Grand Bay Intensive Studies of Atmospheric Mercury

Fong Ngan [1,2,*], **Mark Cohen** [1], **Winston Luke** [1], **Xinrong Ren** [1,2] **and Roland Draxler** [1]

[1] Air Resources Laboratory, National Oceanic and Atmospheric Administration,
5830 University Research Court, College Park, MD 20740, USA;
E-Mails: Mark.Cohen@noaa.gov (M.C.); Winston.Luke@noaa.gov (W.L.);
Xinrong.Ren@noaa.gov (X.R.); Roland.Draxler@noaa.gov (R.D.)

[2] Cooperative Institute for Climate and Satellites, University of Maryland,
5825 University Research Court, College Park, MD 20740, USA

* Author to whom correspondence should be addressed; E-Mail: Fantine.Ngan@noaa.gov

Academic Editor: Robert W. Talbot

Abstract: Measurements at the Grand Bay National Estuarine Research Reserve support a range of research activities aimed at improving the understanding of the atmospheric fate and transport of mercury. Routine monitoring was enhanced by two intensive measurement periods conducted at the site in summer 2010 and spring 2011. Detailed meteorological data are required to properly represent the weather conditions, to determine the transport and dispersion of plumes and to understand the wet and dry deposition of mercury. To describe the mesoscale features that might influence future plume calculations for mercury episodes during the Grand Bay Intensive campaigns, fine-resolution meteorological simulations using the Weather Research and Forecasting (WRF) model were conducted with various initialization and nudging configurations. The WRF simulations with nudging generated reasonable results in comparison with conventional observations in the region and measurements obtained at the Grand Bay site, including surface and sounding data. The grid nudging, together with observational nudging, had a positive effect on wind prediction. However, the nudging of mass fields (temperature and moisture) led to overestimates of precipitation, which may introduce significant inaccuracies if the data were to be used for subsequent atmospheric mercury modeling. The regional flow prediction was also influenced by the reanalysis data used to initialize the WRF simulations. Even with observational nudging, the summer case simulation results in the fine resolution domain inherited features of the reanalysis data, resulting in different regional wind patterns. By contrast, the spring intensive period showed less influence from the reanalysis data.

Keywords: WRF; mercury; nudging; reanalysis data; Grand Bay

---

1. Introduction

Mercury pollution remains a concern because of its effects on ecosystems, including threats to public health through fish consumption [1]. Atmospheric emissions and subsequent deposition are a significant pathway for mercury loading to ecosystems. It is therefore important to understand the sources of atmospheric mercury emissions, as well as its atmospheric fate and transport. The Great Lakes region has been one focus of study [2–4], while mercury pollution in the Gulf of Mexico region has also drawn increasing attention in recent years [5]. The observed atmospheric wet deposition flux of mercury in the Gulf of Mexico region is higher than that in any other region in the United States [6–8]. This is at least partly due to the relatively high rainfall rates in the region, as well as the higher prevalence of tall convective thunderstorms, resulting in the increased scavenging of free tropospheric mercury [9]. The Air Resources Laboratory (ARL) at the National Oceanic and Atmospheric Administration (NOAA) operates a station for the long-term monitoring of atmospheric mercury and other trace species at the Grand Bay National Estuarine Research Reserve (NERR) site in Moss Point, MS (30.412°N, 88.404°W). The station, located about 5 km from the Gulf, was one of the first such sites established in the National Atmospheric Deposition Program's Atmospheric Mercury Network (AMNet). Measurements at the site support a range of research activities aimed at improving the understanding of the atmospheric fate and transport of mercury. Routine monitoring includes speciated Hg, ancillary chemical species and meteorological variables. Two intensive measurement periods were conducted at the site in summer 2010 and spring 2011 [8]. The atmospheric levels of gaseous oxidized mercury (GOM), the form of mercury that is most ecologically important, because it is readily deposited and is more bioavailable once deposited (e.g., most easily methylated), are highest in these two seasons.

Atmospheric models that simulate the fate and transport of emitted mercury can provide comprehensive source attribution information for a given region. The hybrid single-particle Lagrangian integrated trajectory (HYSPLIT) model [10] is a widely used trajectory and dispersion model for estimating source-receptor information on air pollutants. HYSPLIT back-trajectory analyses have been used in numerous atmospheric mercury studies [11–13]. In particular, a special version of the model (HYSPLIT-Hg), with an enhanced treatment of mercury chemistry and phase partitioning in the atmosphere, has been used to estimate detailed source-receptor relationships for atmospheric mercury deposition to the Great Lakes [2]. For such mercury modeling and for the comparable analysis of other pollutants, meteorological data are a critical input, as they control the transport and dispersion of pollutants, as well as the processes of wet and dry deposition. Atmospheric mercury is sensitive to the wind and temperature near the surface, as well as the planetary boundary layer (PBL) condition [14]. Hence, the quality of the meteorological data used influences the accuracy of the model results. The data should properly represent the spatio-temporal variations in the continuous fields of wind, temperature, precipitation, mixing and other relevant atmospheric properties. To better describe the mesoscale features in the region surrounding the NERR site during the measurement intensives, fine-resolution meteorological simulations were conducted by using the Weather Research and Forecasting (WRF) model [15] with data assimilation. As with any such model, errors will inherently exist in WRF-predicted fields due to the uncertainties associated with model dynamics, numerical algorithms, parameterization schemes, model initializations and input data. The inaccuracy of meteorological fields generated by the WRF model will be carried over to any pollutant fate and transport modeling using the data, resulting in errors in the estimates of source-receptor relationships. Thus, a comprehensive evaluation of WRF-generated meteorological fields is necessary before they are used to model the atmospheric transport and deposition of mercury.

To run the WRF model over a regional domain, reanalysis data, which are the results of other models (e.g., global reanalysis), are required to provide the initial conditions (IC) and lateral boundary conditions (LBC) for the simulations. Physical properties, such as momentum, heat and moisture, from the reanalysis data provide the initial state of the atmosphere in the regional domain and constrain the development of weather systems within time-varying lateral boundaries. However, biases in the IC/LBC data can affect the accuracy of the regional simulation results in dynamic downscaling [16]. There is no consensus about which reanalysis data provide the best IC/LBC for regional simulations. IC/LBC data with higher temporal and/or spatial resolution may not necessarily lead to better regional modeling results. The choice of IC/LBC data for the regional model can thus significantly influence the prediction of regional wind patterns, even in the finest resolution domain of a three-nested-domain configuration [17]. Thus, WRF performance must be assessed by using alternative IC/LBC inputs.

For reducing model bias during the simulation, four-dimensional data assimilation (FDDA), or nudging, is a well-known and efficient method available in the WRF model [18]. This approach adjusts model values toward observations (observational nudging) or toward analysis fields (grid or analysis nudging) for temperature, moisture and the u- and v-components of wind at each integration time step. Furthermore, Deng *et al.* [18] also suggested that in order to provide more accurate IC/LBC data for the regional modeling, reanalysis data must undertake an objective analysis process with surface and upper level meteorological data before being used to initialize the WRF model. Studies [17,19–23] have demonstrated that nudged results outperform simulations without nudging, which is also beneficial for air quality applications, including chemical and dispersion modeling. However, there are different suggestions for configuring the nudging process in order to minimize model bias, and the situation is complex, as physics schemes for the surface and PBL develop their own local-scale dynamics regardless of the influence of "artificial" nudging. Grid nudging of wind within the PBL (including the surface layer) is the preferred protocol of the U.S. EPA for retrospective meteorological simulations used in air quality studies [24]. Rogers *et al.* [22] presented sensitivity tests for nudging configurations and found that multiscale FDDA combining both analysis and observational nudging produced the smallest errors. Hegarty *et al.* [23] found that grid nudging within the PBL corrected an overestimate in plume transport possibly caused by a positive surface wind bias in WRF. Godowitch *et al.* [21] showed that eliminating nudging within the surface layer and PBL resulted in a better prediction of nocturnal jet speed. Gilliam *et al.* [24] suggested that the use of the wind profile aloft without surface nudging improved modeled winds within the convective PBL and above the stable PBL at night.

The overall goal of this study is to find a suitable WRF-model configuration to develop a high-quality, high-resolution regional- and local-scale meteorological dataset to support mercury modeling for the two measurement campaigns at the Grand Bay station conducted in summer 2010 and spring 2011. We evaluate the WRF model performance through comparison with observations, including surface measurements and soundings at the Grand Bay NERR station. WRF simulations are generated by using different reanalysis data for IC/LBC to examine influences on the prediction of regional flows in the WRF model. Another set of WRF simulations is conducted to examine the effect of the nudging strategy on model performance. Through the evaluation of meteorological model results, we may be able to identify the inaccuracy in meteorological data affecting the simulation of pollutant transport. Section 2 describes the model configuration for WRF, experiment designs (*i.e.*, the WRF model configurations) for the different sets of simulations carried out, sources of observation, model evaluation methodology and the HYSPLIT model used for the backward trajectory analysis, as well as reviews the two Grand Bay Intensive campaigns, including synoptic weather and mercury episodes. In Section 3, the model results and analyses are presented for the summer and spring campaigns. Our conclusions are presented in Section 4.

2. Experimental Section

2.1. Configuration of the WRF Model

The Advanced Research WRF version 3.2 is used to simulate weather conditions during two Grand Bay Intensive measurement periods in summer 2010 (28 July to 15 August) and spring 2011 (19 April to 9 May). The modeling domains (Figure 1) include three nested grids with 36-km resolution (D01) for the contiguous U.S., 12-km resolution (D02) for the Eastern U.S. and 4-km resolution (D03) for the Gulf Coast of Louisiana, Mississippi and Alabama. Altogether, 43 vertical sigma layers are used, with the higher resolution near the ground to better describe the atmospheric structure in the lower boundary layer. The thickness of the lowest layer is around 34, and 15 layers are included below 850 hPa (1.5 km). The two coarser domains are run with two-way nesting, while the finest domain (D03) is initialized by using the D02 result. The physics options include the rapid radiative transfer model for longwave radiation [25], the Dudhia scheme for shortwave radiation [26], the Pleim–Xiu land surface model [27,28], the Asymmetrical Convective Model 2 for PBL parameterization [29], WSM3 for microphysics [26] and the Grell–Devenyi Ensemble [30] for the sub-grid cloud scheme.

2.2. WRF Simulation Designs

2.2.1. Reanalysis Data for WRF Model Initialization

To initialize the WRF model for the North American regional domain, either reanalysis data or global model products can provide the required meteorological fields for IC/LBC. Four reanalysis products are used to initialize the WRF simulations for the Grand Bay Intensive period of summer 2010:

(1) the National Centers for Environmental Prediction (NCEP) North American Regional Reanalysis (NARR) at 32-km resolution with 29 pressure levels [31]; (2) the NCEP Global Final Analysis with data assimilation from the Global Forecast System (GFS) in 1 × 1 degree spacing and 26 vertical layers [32]; (3) the NCEP-National Center for Atmospheric Research Reanalysis Product (NNRP) in a 2.5 degree latitude-longitude grid with 17 pressure levels and 28 sigma levels [33]; and (4) the NCEP Climate Forecast System Reanalysis (CFSR) at 38-km spatial resolution and 64 pressure levels [34]. They are all available at 6-h intervals; except NARR, which is 3 hourly. We evaluate the results of the WRF modeling run by using the same model configuration (detail shown in Section 2.1), but with the four sources of IC/LBC data above.

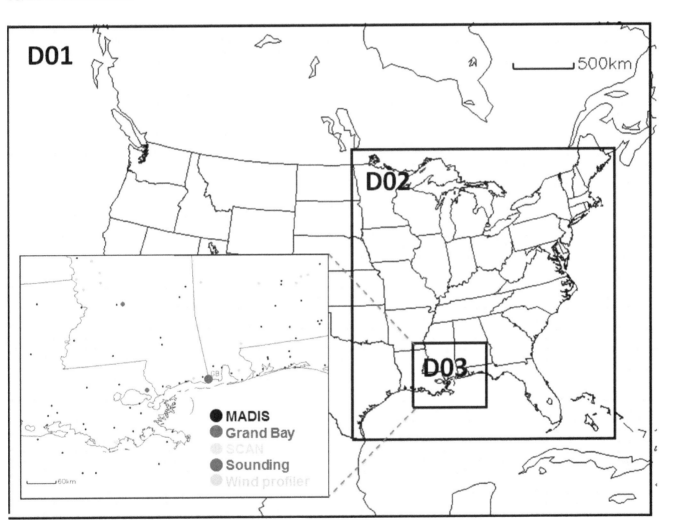

Figure 1. The three nested domains for WRF simulations and observation stations used in this study.

2.2.2. Nudging Procedure

The practice of objective analysis together with grid and observational nudging in meteorological modeling has been shown to be beneficial in generating inputs for air quality studies. Objective analysis modifies the first guess (*i.e.*, reanalysis data) by ingesting information from given observations to provide more accurate IC/LBC data for WRF initialization. Objectively analyzed data also provide analysis fields for the grid nudging of wind, temperature and moisture to constrain the error growth during the simulation. In this study, the surface, sounding and wind profiler data obtained from the Meteorological Assimilation Data Ingest System (MADIS) are ingested in the simulations through objective analysis and nudging to maintain the model results close to the given observations. The Grand Bay measurements and the Soil Climate Analysis Network (SCAN) data are not included in the objective analysis and nudging, but rather reserved as an independent dataset for the model evaluation. Grid nudging and/or observational nudging are applied for all of the domains throughout the simulations to minimize errors. Three WRF simulations with different nudging configurations are conducted for the summer 2010 campaign: (1) allDA, grid nudging (including surface) and observational nudging of all fields, temperature, moisture and the u- and v-components of wind; (2) wdDAno3D, observational nudging of wind components, but no grid nudging; and (3) wdDA, grid nudging (including surface) and the observational nudging of wind fields.

2.3. Observation Data for Model Evaluations

The locations of observation data used in the analysis are shown in Figure 1. This study uses meteorological measurements from three sources: (1) The Grand Bay NERR site provides surface meteorological measurements and extra soundings launched during the intensive period. Details on the Grand Bay station and measurements taken during the two campaigns are contained in [8]. (2) The MADIS (http://madis.noaa.gov/) developed by the NOAA Earth System Research Laboratory collects observations from a wide range of national and international data sources, and its data are integrated and quality controlled. Relevant surface and upper level (sounding and wind profiler) observations are downloaded and extracted for the study domains. (3) The SCAN (http://www.wcc.nrcs.usda.gov/scan) operated by the Natural Resources Conservation Service provides regular meteorological observations, including temperature, relative humidity, pressure, wind speed and wind direction. Two SCAN sites are relatively close to the Grand Bay station, while the rest are inland.

The evaluation of WRF results focuses on the 4-km domain (D03) for wind speed, wind direction, temperature and relative humidity. Gridded model values are paired with corresponding observations in space and in time for the comparison. If a missing data point was encountered, the station at that particular time step would be skipped. To assess the model performance, we provided domain-wide statistics calculated with simulated and measured values. a combination of metrics is suggested for model evaluations, since a single statistical metric only provides a limited characterization of the errors [35]. Statistical summaries include: the correlation coefficient (R) describing the extent of the linear relationship between the model and observed values; the mean absolute error (MAE) and root mean square error (RMSE), together with mean bias (bias), summarizing how close the predicted values are to the measured values; the standard deviation of the error (SDE), measuring the amount of variation from the average error; Index of Agreement (IOA), indicating how well the model represents the pattern of perturbation about a mean value. These statistical measures are commonly used for evaluations of meteorological and air quality models ([20,36,37], etc.), and a more detailed description of them can be found in [38].

2.4. Backward Trajectory Analysis

The ultimate use of the WRF fine-resolution meteorological results is to support the trajectory and dispersion analysis of atmospheric mercury in the Grand Bay region. The HYSPLIT model is used to understand how different meteorological inputs affect the estimates of source-receptor relationships. The model is designed for both simple air parcel trajectories and complex dispersion and deposition simulations. It has been used for trajectory analysis to identify the source-receptor relationships of air pollutants and for dispersion predictions for a variety of events, such as nuclear incidents, volcanic eruptions, wild fire smoke transport and dust storm episodes (http://ready.arl.noaa.gov/index.php). Backward trajectories are computed for particular mercury episodes during the intensive period utilizing WRF meteorology conducted with different model configurations, as described in the previous sections. Trajectory analysis can be a useful assessment tool for selecting more accurate WRF meteorology and estimating meteorological uncertainty, because the trajectories represent the space- and time-integrations of the velocity fields rather than just relying on verification statistics at fixed locations.

2.5. Overview of Grand Bay Intensive Measurements Periods

For the Grand Bay Intensive measurement in summer 2010 (28 July to 15 August) a high pressure system was dominant in the Gulf at the beginning of the study period. The system was weakening, and a weak stationary front was approaching the coast of Mississippi on 2 August. During this period, the ambient temperature was around 33 °C as the daytime maximum and 25 °C as the nighttime minimum, while the wind speed was light to moderate with a mostly southerly to southwesterly flow. Gaseous oxidized mercury (GOM) mid-day peaks of 20–60 pg/m^3 were observed at Grand Bay on 2, 4–7 August (indicated in Figure 2, top panel), while the typical peak value ranged about 10–20 pg/m^3. The average gaseous elemental mercury (GEM) level of the campaign period was 1.42 ng/m^3 [8]. Another stationary front approached the Grand Bay site on 7 and 8 August. Moisture-rich Gulf air interacting with the approaching stationary front led to thunderstorm development, and a significant amount of rainfall was measured at Grand Bay during this period.

Another Grand Bay Intensive period was conducted from mid-April to mid-May 2011 (19 April to 9 May). During the first 9 days, frontal activities were confined largely to the north of the Grand Bay region, and the area was dominated by southerly flows. On 28 April and 3 May, cold fronts passed through the area, bringing continental cold air masses to the station. After these frontal passages, the area experienced post-frontal conditions, namely dry air, low night-time temperatures (as low as 10 °C), a light northeasterly wind in the morning and a southerly sea breeze in the afternoon. Wind speed was moderate in general, but became gradually smaller after the second frontal passage on 3 May. Mercury episodes were identified on 29 April and 4–7 May (indicated in Figure 7, top panel), as high levels of GOM were observed at the Grand Bay site with peak values of 30–70 pg/m^3 [8].

3. Results and Discussion

3.1. Meteorological Modeling for Summer 2010

3.1.1. Regional Evaluations

The statistical scores for each simulation were first computed by using MADIS surface data, which were also assimilated in the observational nudging, and then SCAN data, which were not used in the nudging. a comparison against the data from the Grand Bay NERR site is presented in the next section. Table 1 shows a comparison of the model performance by using the three different nudging configurations. It can be seen that the wdDA case (grid and observational nudging only for the u- and v-components of wind), outperformed the other two in predicting wind speed. The AllDA case, which used all nudging for wind, temperature and moisture, also had relatively good results, while the simulation without 3D grid nudging (wdDAno3D) had the worst statistical scores. Hence, the 3D grid nudging appears to improve wind predictions in this situation. Since allDA nudged both mass fields (temperature and moisture) and wind fields, perhaps not surprisingly, it exhibited better scores in temperature prediction. Table 2 compares MAE statistics by using the independent (*i.e.*, not included in the objective analysis and nudging) SCAN observational data. It is evident that wdDA demonstrated the best skill at simulating wind speed and direction. The inclusion of mass fields in the nudging or removal of grid nudging degraded the accuracy of wind prediction in this application. For temperature and relative humidity, the MAE of the wdDAno3D simulation was lowest of all.

Table 1 also presents the model performance statistics (based on MADIS surface data) among cases using the four different reanalysis datasets for model initialization. For this comparison, the wdDA nudging scheme was used in each case. The best statistical scores for wind speed and direction were exhibited by the GFS case, while for surface temperature, the CFSR result was slightly better than the others. Note that the reanalysis data provide IC/LBC, as well as analysis fields for 3D grid nudging. The comparison with independent SCAN data in Table 2 shows that the NARR-based simulation was more accurate at predicting wind speed. For wind direction, the MAE for the CFSR-based simulation was slightly lower than the others, while for temperature and relative humidity, the GFS-based simulation exhibited the best performance.

Table 1. The statistical summary of the D03 domain for the summer 2010 intensive period computed by using the surface sites (METAR) from the Meteorological Assimilation Data Ingest System (MADIS). IC, initial conditions; LBC, lateral boundary conditions; SDE, standard deviation of the error; IOA, Index of Agreement; NARR, North American Regional Reanalysis; GFS, Global Forecast System; NNRP, NCEP-National Center for Atmospheric Research Reanalysis Product; CFSR, Climate Forecast System Reanalysis.

Variable	IC/LBC	Nudging	R	Bias	RMSE	MAE	SDE	IOA
Wind speed	WRF-NARR	allDA	0.684	−0.195	1.127	0.842	1.285	0.819
$(m \cdot s^{-1})$	WRF-NARR	wdDAno3D	0.617	0.022	1.222	0.938	1.554	0.783
17,447 samples	WRF-NARR	wdDA	0.716	−0.223	1.049	0.797	1.176	0.831

Table 1. *Cont.*

Variable	IC/LBC	Nudging	R	Bias	RMSE	MAE	SDE	IOA
	WRF-GFS	wdDA	0.756	−0.339	1.009	0.757	1.038	0.842
	WRF-NNRP	wdDA	0.721	−0.340	1.069	0.799	1.113	0.821
	WRF-CFSR	wdDA	0.738	−0.333	1.037	0.777	1.078	0.831
Wind direction	WRF-NARR	allDA	0.719	−7.396	70.223	36.123	75.511	0.850
(degree)	WRF-NARR	wdDAno3D	0.665	−6.122	76.290	42.114	84.132	0.821
16,247 samples	WRF-NARR	wdDA	0.731	−5.565	68.692	35.011	74.529	0.857
	WRF-GFS	wdDA	0.765	−5.504	65.571	30.246	69.867	0.878
	WRF-NNRP	wdDA	0.729	−2.319	68.984	33.354	75.608	0.858
	WRF-CFSR	wdDA	0.744	−5.693	67.782	32.612	72.710	0.866
Temperature	WRF-NARR	allDA	0.940	−0.097	1.225	0.869	1.444	0.966
(°C)	WRF-NARR	wdDAno3D	0.830	−0.212	1.992	1.505	2.366	0.901
25,585 samples	WRF-NARR	wdDA	0.850	0.093	1.871	1.361	2.368	0.915
	WRF-GFS	wdDA	0.857	−0.356	1.872	1.410	2.119	0.919
	WRF-NNRP	wdDA	0.853	0.171	1.864	1.353	2.401	0.853
	WRF-CFSR	wdDA	0.864	−0.226	1.804	1.347	2.112	0.923

Table 2. The MAE of the D03 domain for the summer 2010 intensive period computed by using the surface data from Soil Climate Analysis Network (SCAN).

ICBC	Nudging	Wind Speed (m·s^{-1}) 3806 samples	Wind Direction (degree) 3961 samples	Temperature (°C) 4158 samples	Relative Humidity (%) 4173 samples
WRF-NARR	allDA	1.180	61.425	2.230	8.860
WRF-NARR	wdDAno3D	1.222	60.476	1.858	7.663
WRF-NARR	wdDA	1.171	59.629	2.482	8.912
WRF-GFS	wdDA	1.251	60.680	1.651	8.334
WRF-NNRP	wdDA	1.366	61.566	1.787	9.246
WRF-CFSR	wdDA	1.207	58.772	2.021	8.806

3.1.2. Grand Bay Station Analysis

The meteorological observations at the Grand Bay site were not included in the data assimilation, similar to the SCAN data, meaning that they can be used as an independent dataset for the model evaluation. Table 3 shows a statistical summary (MAE) of model performance at this site based on a comparison with four soundings launched during the 2010 intensive. It can be seen that the influences of the reanalysis data through IC/LBC (and nudging) reached the finest domain (D03). Similar wind speed errors in the upper atmosphere were generated by the three nudging configurations. However, when grid nudging was turned off (wdDAno3D), larger MAEs were produced for wind direction, temperature and relative humidity. For this run, only observational nudging was operating, with the available data at the surface and aloft, including two soundings and one wind profiler in the study domain. The grid nudging used in the other configurations appears to have had a positive impact in the upper atmosphere on restraining the error growth. Among the four cases using different reanalysis data, the MAE scores of the GFS case were the lowest for wind prediction, except for relative humidity.

Table 3. The MAE of the D03 domain for the summer 2010 intensive period computed by using four soundings launched at Grand Bay.

ICBC	Nudging	Wind Speed $(m \cdot s^{-1})$ 392 samples	Wind Direction (degree) 392 samples	Temperature (°C) 392 samples	Relative Humidity (%) 392 samples
WRF-NARR	allDA	1.641	33.274	0.636	10.854
WRF-NARR	wdDAno3D	1.656	41.510	0.780	14.003
WRF-NARR	wdDA	1.613	32.391	0.652	11.254
WRF-GFS	wdDA	1.548	31.722	0.607	15.003
WRF-NNRP	wdDA	2.054	33.469	0.731	9.671
WRF-CFSR	wdDA	1.898	30.434	0.603	13.425

WRF showed a good prediction for the 2-m temperature at the Grand Bay site during the daytime, but had a warm bias occasionally at nighttime (Figure 2, middle panel). The bottom panel of Figure 2 shows a time series of the modeled and observed precipitation at the Grand Bay site. There are several periods of modeled precipitation with no corresponding observations and several periods of observed precipitation without corresponding model predictions. Usually, a rapid drop in 2-m temperature associated with rain can be observed. The model missed the rainfall event on 3 August and did not predict the accurate timing of rain on 8 and 9 August. The Gulf area in the summer has abundant moisture available, and thunderstorms can easily be triggered. However, it is debatable if sub-grid-scale cloud parameterization should be used at 5–10-km resolutions, while at the same time, modeling using explicit cumulus schemes has not yet been successful [17,18]. Sensitivity tests on nudging configurations were carried out to determine if the model's precipitation performance could be improved (Figure 2, lower panel). When temperature and moisture were nudged through grid and observational nudging (allDA), the model generated the most significant overestimates of precipitation. For daily accumulated rainfall over the entire D03 domain, the allDA nudging scheme produced 3–5-times more precipitation than the other two cases (figure not shown). Among all MADIS surface stations, we picked up sites where zero precipitations was recorded and corresponding model values for the comparison in Figure 3. It is noted that allDA has the largest overestimate of precipitation compared with the other two simulations using different nudging configurations. As some pollutants (e.g., GOM) are highly vulnerable to wet removal processes, extraneous precipitation would lead to artificially high wet deposition. Turning off the observational nudging for temperature and moisture (wdDA and wdDAno3D) reduced the precipitation overestimates somewhat. As can be seen from the top panel in Figure 2, these two nudging schemes did not produce large overestimates of wind speed at the Grand Bay site.

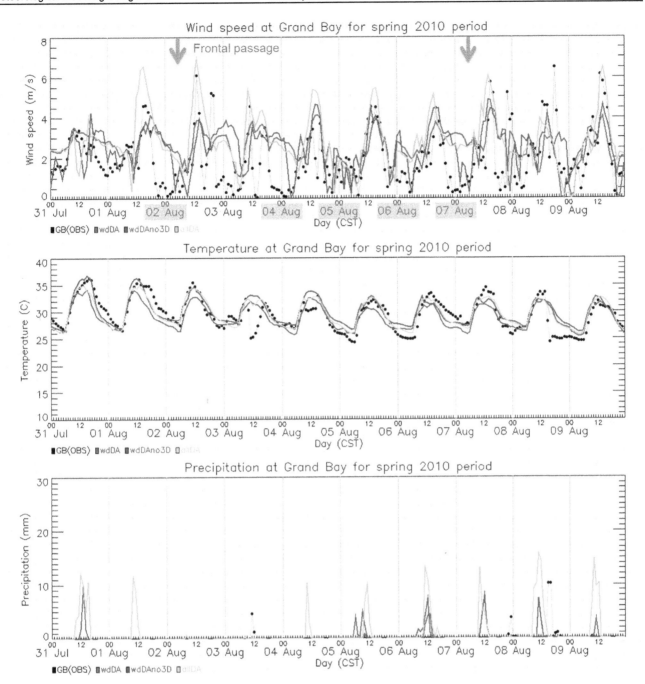

Figure 2. Time series of observed (dot-gray line) and modeled (various colors for different nudging configurations) (**top**) 10-m wind speed (m·s^{-1}), (**middle**) 2-m temperature and (**bottom**) hourly precipitation (mm) at the Grand Bay site. Gray arrows indicate frontal passages, and mercury episodic days are highlighted with gray boxes.

The surface wind comparisons in Figure 2 show that the model failed to predict the decrease in surface wind speed at night on certain days during the simulation period. This has been reported in other studies, such as those in Southeastern Texas [39,40] and the coastal cities of Spain [41]. On those days, a high pressure system was dominant over the study area and no precipitation was predicted, as shown in Figure 2. Possible causes of the overestimation of nocturnal wind speed could be the excessive vertical mixing simulated by the PBL scheme [39,40], the decoupling of the nocturnal boundary layer not properly simulated in the model [42] and/or the inaccuracy on predicting surface flux in the surface parameterization [43]. The wind rose plots (Figure 4) are the observed and model wind distribution at the Grand Bay site during the simulation period. All of the simulations generated too much south-southwesterly wind, while the measurement at Grand Bay was north-northwesterly dominant during the campaign period. For the wind speed, the observation had small values for northerly, but large values for southwesterly flows. The model re-produced strong southerly components of wind, but failed to generate a calm northerly. Examining the wind rose plots generated from the reanalysis data (not shown), we noticed that the reanalysis data that were input for WRF simulation lacked northerly components, while the wind measurement at Grand Bay station showed that the dominant wind was northerly. The NARR reanalysis had almost no northerly wind simulated, and the predominant wind directions were southerly to southwesterly. The model inherited features of the NARR reanalysis data, including the missing northerly components for the wind prediction.

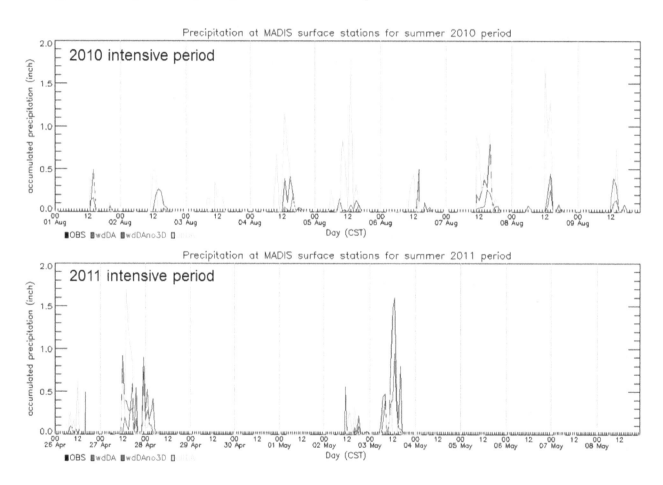

Figure 3. Time series of hourly precipitation (inches) total accumulation over MADIS surface stations with zero precipitation for 2010 intensive (**top**) and 2011 intensive (**bottom**).

3.1.3. Backward Trajectory Analysis

On 4 August at 20:30 UTC, the GOM level reached 62 pg/m^3 at the Grand Bay station, one of the highest measured GOM concentrations during the summer 2010 intensive period. During this peak, the measured surface wind direction at the site was southwesterly to westerly, while the wind speed was moderate, about 4 m·s^{-1}. Backward trajectories (Figure 5) ending at 21:00 UTC on 4 August

2010, at Grand Bay were computed by HYSPLIT utilizing meteorological fields from WRF simulations initialized with different reanalysis data. Four starting altitudes, expressed as a fraction of the local model-estimated PBL height, were chosen: 0.05, 0.3, 0.5 and 0.95. The GFS-, NNRP- and CFSR-based simulations showed air parcels arriving at the site from the west, potentially bringing pollutants from sources in the west to Grand Bay. However, the NARR-based simulation indicated air masses coming from the Gulf, where the air would be expected to be relatively clean. The time series of the wind profile at Grand Bay (figure not shown) further showed that the NARR-based simulation predicted stronger and more southerly near surface winds compared with the GFS-based simulation. In this study, observations, including surface, profiler data and sounding, were used to adjust the reanalysis data (used for IC/BC and grid nudging) toward the observations through objective analysis. In addition, observations were directly used to nudge the predicted values during the simulation. With all of this "forcing" by observations and even with the three-nested domain configuration, the finest resolution domain still inherited differing features of reanalysis data, which resulted in different model-predicted regional wind patterns. The same backward trajectory configuration was applied for nudging cases (Figure 6). The wdDAno3D-based meteorology generated quite different backward trajectories when they were initiated at higher elevations (0.5 and 0.95 of the local model-estimated PBL height). Since the evaluation based on the Grand Bay sounding showed a larger error in wind direction prediction than the other cases, the northerly backward trajectories in the wdDAno3D case are likely to be unrealistic.

3.2. Meteorological Modeling for Spring 2011

3.2.1. Regional Evaluations

For the spring 2011 intensive period, all three nudging configurations conducted for the summer 2010 case were tested for the spring period, allowing us to further examine their performance in different seasons and meteorological conditions. To understand the variations in the reanalysis data, we focus efforts on evaluating the use of NARR and GFS datasets for modeling the spring episode. These are commonly used for the initialization of regional models in the community, and the benefit of using the other two datasets (NNRP and CFSR) for IC/LBC is not evident in the summer case.

Both simulations generated similar results for surface wind and temperature compared with the MADIS data. Consistent with the comparison with MADIS data for the summer 2010 intensive period, the nudging performed during the simulation successfully adjusted the predicted values toward the given observations. Table 4 shows the statistical summary of the model results evaluated against the independent SCAN data (*i.e.*, data that were not used in the nudging). For mass fields (temperature and relative humidity), the NARR and GFS runs had similar statistical scores, while the NARR-based simulation performed slightly better for predicting wind speed and wind direction. Both wdDA and allDA had better scores for wind prediction, implying that the grid nudging for wind does help reduce the error for wind fields. The NARR case did generate wind fields as good as GFS in contrast to the summer case, for which NARR failed to re-produce southerly winds, as the GFS case did. The model over-predicted the amount of precipitation for a few days of the spring episode (Figure 3). AllDA had more days of rainfall overestimation than the other two cases, but overall, it was not as severe as the summer period.

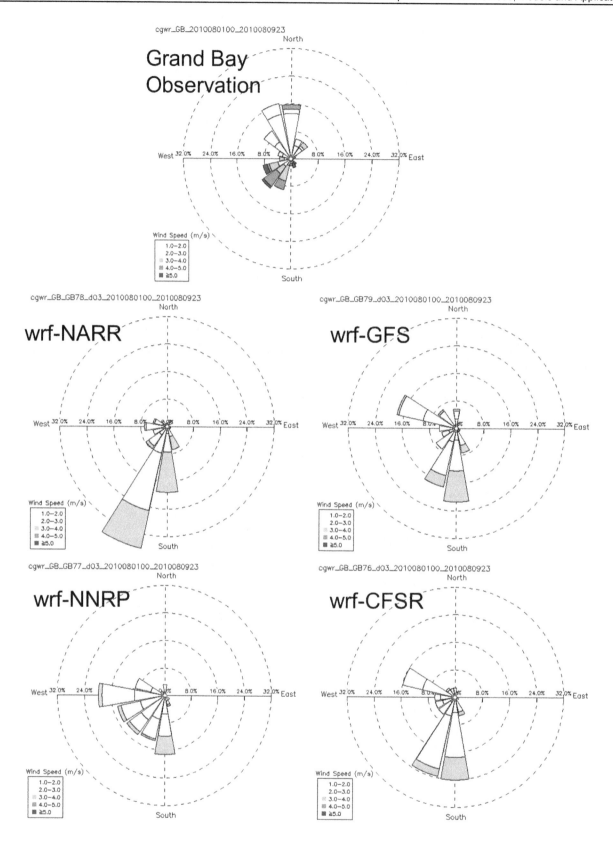

Figure 4. Wind rose plots at Grand Bay station for four simulations using different reanalysis data for WRF initialization during the simulation period of the summer 2010 campaign.

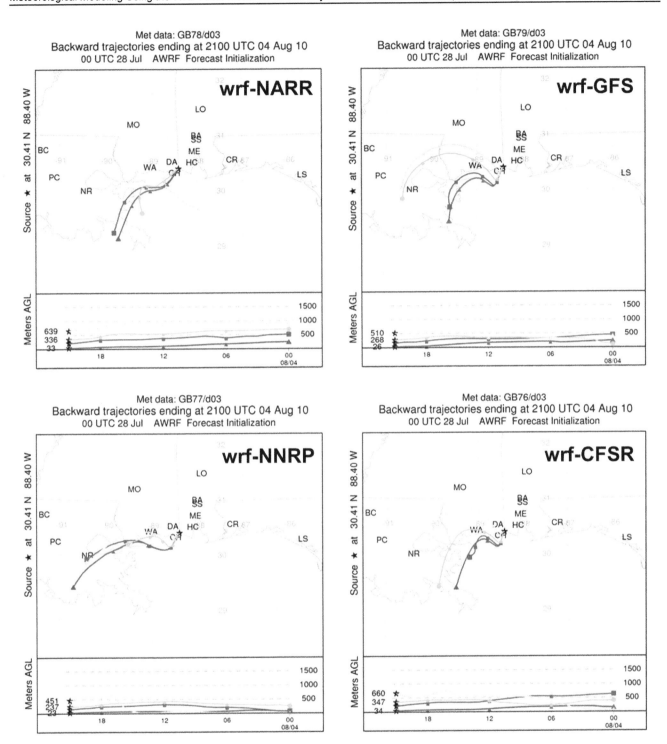

Figure 5. Backward trajectories ending at 21:00 UTC on 4 August 2010, at Grand Bay at multiple heights utilizing meteorological inputs from the NARR, GFS, NNRP and CFSR cases. Names in each plot indicate potential sources of mercury emissions.

3.2.2. Grand Bay Station Analysis

As shown in the previous section with the summer intensive period, statistical summaries can only reveal the model's performance in a general way. The examination of the time series of the meteorological variables at an individual site can yield further insights. Figure 6 shows the time series of 10-m wind speed, wind direction and the 2-m temperature at the Grand Bay site during the spring 2011 intensive period. Both the NARR- and GFS-based simulations predicted similar wind patterns, generally capturing the turning of the wind direction after the frontal passages on 28 April 28 and 3 May, as well as the dominant southerly flow between 30 April and 2 May. The NARR case did not appear to predict anomalous southerly winds at the Grand Bay site as it had for portions of the summer 2010 intensive period. The NARR-based simulation generally predicted a larger diurnal variation of 2-m temperature than the GFS case. This confirms the statistical scores that the surface temperature error was larger in the spring period than in the summer episode. Surface temperature may affect the prediction of the PBL height, but it is not a particularly influential parameter for dispersion modeling. Therefore, the inaccuracy of modeling the 2-m temperature is not expected to significantly affect the accuracy of the atmospheric mercury simulations using these data. The spring 2011 intensive period was less stormy than the earlier intensive; rain occurred before two frontal passages, but it did not last long. No precipitation was observed at Grand Bay on days with high atmospheric GOM concentrations (highlighted with gray boxes in Figure 7), and this lack of precipitation at the site was simulated well by the modeling.

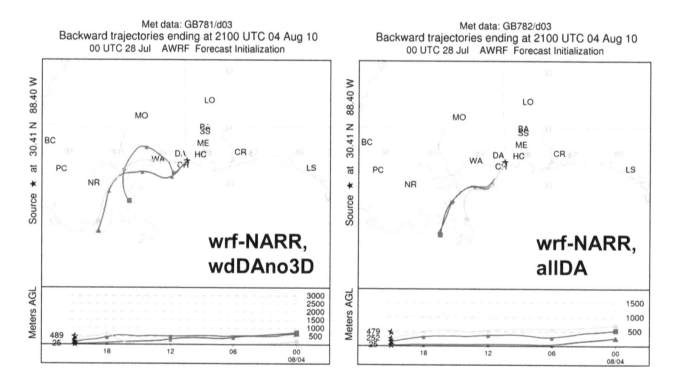

Figure 6. Backward trajectories ending at 21:00 UTC on 4 August 2010, at Grand Bay at multiple heights utilizing meteorological inputs from the wdDAno3D and allDA cases. Names in each plot indicate potential sources of mercury emissions.

The regional wind pattern is a critical factor in determining the transport of atmospheric mercury and other pollutants. Figure 8 shows the wind rose plots for surface wind over the entire innermost domain (D03) during the spring 2011 campaign. The surface wind distribution in the simulations with different initialization data was similar, in contrast with the modeling results for the summer 2010 intensive. The wind patterns from the reanalysis data represented closely matched the measurements at the Grand Bay station. Even though northerly components were somewhat under-represented, both NARR and GFS reanalysis reproduced two of the main wind patterns (northerly wind and southeasterly wind), as shown in the measurement. This finding suggests that the influence of the reanalysis data over the regional flow prediction is not always significant, but depends on the atmospheric conditions. Table 5 shows the model error (MAE) in the upper levels using data from the 10 soundings collected at Grand Bay during the spring 2011 intensive. All four variables had very similar MAE scores among the simulations with different model configurations. In the comparison of two intensive periods, the reanalysis data had a more obvious impact on WRF performance, even with regard to the innermost domain, in the summer campaign than the one conducted in spring.

Table 4. The statistical summary of the D03 domain for the spring 2011 intensive period computed by using SCAN data.

Variable	IC/LBC	Nudging	R	Bias	RMSE	MAE	SDE	IOA
Wind speed	WRF-NARR	allDA	0.723	−0.268	2.068	1.565	2.426	0.835
$(m \cdot s^{-1})$	WRF-NARR	wdDAno3D	0.725	−0.407	2.070	1.571	2.340	0.819
2768 samples	WRF-NARR	wdDA	0.738	−0.401	2.030	1.546	2.296	0.832
	WRF-GFS	wdDA	0.656	−1.339	2.617	1.966	2.335	0.627
Wind direction	WRF-NARR	allDA	0.513	2.688	75.448	35.057	84.320	0.744
(degree)	WRF-NARR	wdDAno3D	0.400	1.244	85.365	41.831	95.609	0.683
2834 samples	WRF-NARR	wdDA	0.502	1.477	76.431	35.584	84.930	0.737
	WRF-GFS	wdDA	0.471	8.029	80.530	39.094	92.958	0.718
Temperature	WRF-NARR	allDA	0.922	0.466	2.262	1.816	3.179	0.957
(°C)	WRF-NARR	wdDAno3D	0.904	0.280	2.309	1.841	3.123	0.948
2844 samples	WRF-NARR	wdDA	0.910	0.433	2.296	1.855	3.212	0.951
	WRF-GFS	wdDA	0.906	−0.137	2.276	1.759	2.791	0.947
RH	WRF-NARR	allDA	0.880	2.475	9.676	7.683	13.808	0.933
(%)	WRF-NARR	wdDAno3D	0.838	−0.389	12.237	9.520	15.263	0.908
2840 samples	WRF-NARR	wdDA	0.847	−1.357	12.399	9.649	14.855	0.909
	WRF-GFS	wdDA	0.837	4.237	11.879	9.414	17.593	0.901

Table 5. The MAE of the D03 domain for the spring 2011 intensive period computed by using the 10 soundings launched at Grand Bay.

ICBC	Nudging	Wind Speed $(m \cdot s^{-1})$ 978 samples	Wind Direction (degree) 978 samples	Temperature (°C) 978 samples	Relative Humidity (%) 917 samples
WRF-NARR	allDA	1.698	21.938	0.915	9.797
WRF-NARR	wdDAno3D	1.869	21.822	0.895	8.825
WRF-NARR	wdDA	1.683	22.059	0.838	8.575
WRF-GFS	wdDA	1.649	20.128	0.626	8.432

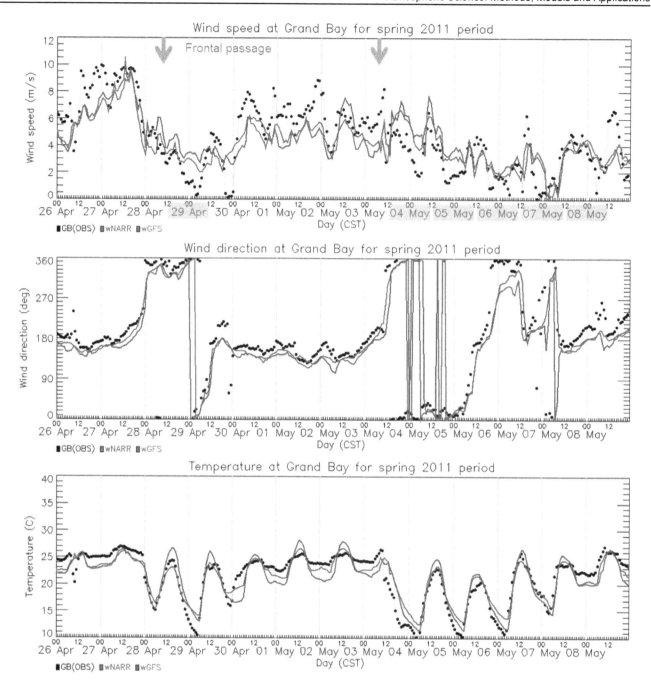

Figure 7. Time series of observed (dot-gray line) and modeled (initialization with NARR and GFS data) (**top**) 10-m wind speed (m·s^{-1}), (**middle**) 10-m wind direction and (**bottom**) 2-m temperature at the Grand Bay site. Gray arrows indicate frontal passages, and mercury episodic days are highlighted with gray boxes.

3.2.3. Backward Trajectory Analysis

Backward trajectories were constructed by using the WRF meteorology results for air masses arriving at the Grand Bay site at 23:00 UTC on 6 May 2011 (Figure 9), when one of the highest GOM measurements (43.50 pg/m^3) during the spring 2011 spring campaign occurred at 22:30 UTC of that day. Both simulations, using NARR and GFS for initialization, indicated that the air came from the west, northwest and/or north, passing in the vicinity of one or more emission sources that may have contributed to the peak. However, the trajectories using the GFS-based WRF results traveled longer distances than those generated by the NARR-based WRF results. The differences in wind speed and direction shown in the trajectory results would likely affect the dispersion modeling results from the relevant sources in the region. Although the wind roses were similar, the trajectories clearly showed persistent differences. Future work will examine the impacts of the use of different WRF model results on mercury dispersion in the vicinity of the Grand Bay site during the intensive periods.

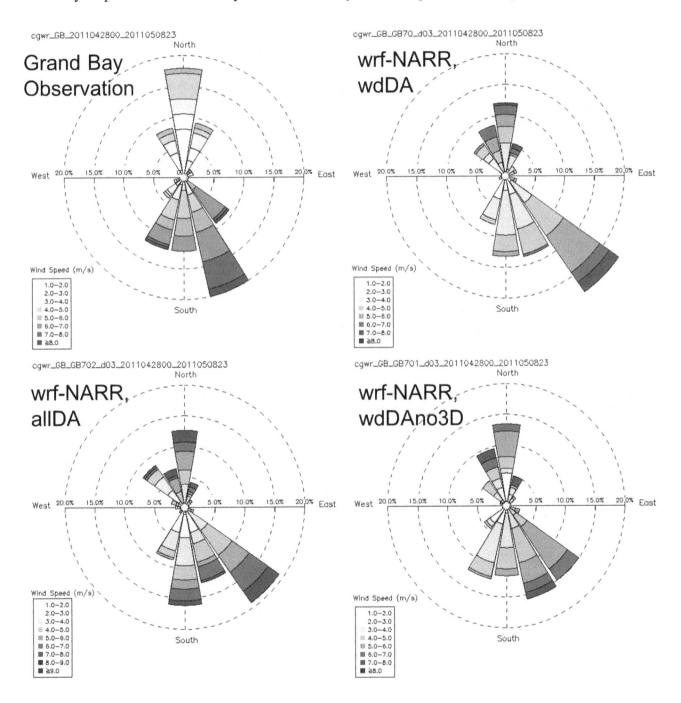

Figure 8. Wind rose plots at Grand Bay for three simulations with different nudging configurations during the simulation period of the spring 2011 campaign.

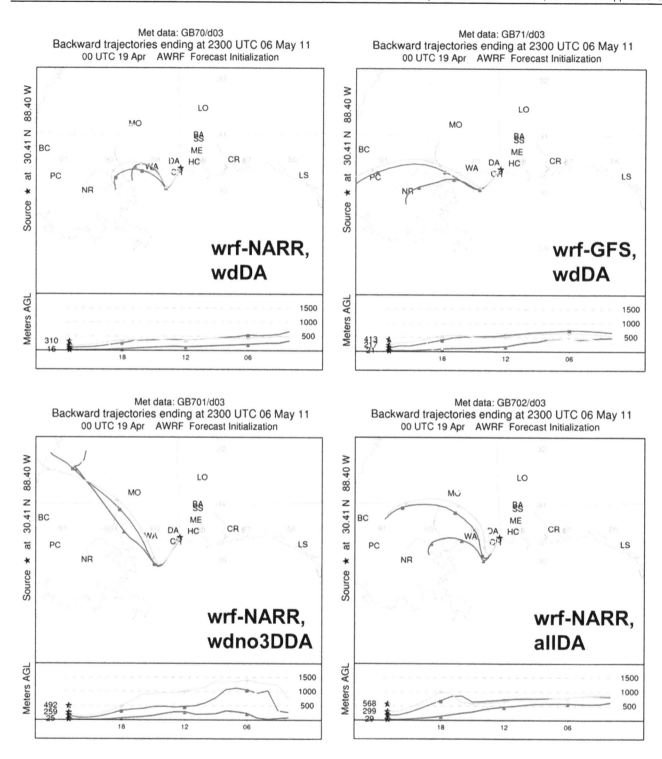

Figure 9. Backward trajectories ending at 23:00 UTC on 6 May 2011, at Grand Bay at multiple heights utilizing meteorological inputs generated by different configurations. The names in each plot indicate potential sources of mercury emissions.

4. Conclusions

A site at the Grand Bay NERR station has been operated by the NOAA's Air Resources Laboratory for the long-term monitoring of atmospheric mercury and other trace species since 2006. Two intensive measurement periods were conducted at the site in summer 2010 (from 28 July to 15 August) and spring 2011 (from 19 April to 9 May) to improve the understanding of the atmospheric fate and transport of mercury. To support mercury modeling in conjunction with the intensive, WRF-ARW was used to develop fine-resolution meteorological fields for the two campaign periods. Two sets of sensitivity tests were performed, to examine the influences on model performance and regional flow predictions: (1) the use of different reanalysis data for WRF initialization; and (2) the use of different nudging configurations. WRF simulations were evaluated with conventional observations and additional measurements, including surface and sounding data obtained at the Grand Bay station during the intensives. Backward trajectories using HYSPLIT were constructed for illustrative mercury peaks with different WRF meteorology to understand the influence of different meteorology inputs on the model-estimated source-receptor relationships at the site.

The nudging process in WRF sufficiently adjusted the model values toward the observations or analysis fields. The simulations by WRF with grid and observational nudging generated reasonable results and were in good agreement with the Grand Bay measurements. It was found that 3D grid nudging at the fine spatial grid (4-km resolution in this study), namely bringing the reanalysis data into the fine spatial grid, did not degrade model performance, but reduced errors in the wind predictions at the surface and aloft (the allDA and wdDA cases). The nudging of mass fields (temperature and moisture) had a significant impact on model-predicted precipitation, especially for the summer 2010 intensive period. In this case, mass-field nudging resulted in more extraneous precipitation at the Grand Bay site and 3–5-times more precipitation in the study domain than that generated in the simulations with the nudging of only wind components. These significant differences in modeled precipitation may have potentially large impacts on mercury fate and transport modeling, through the effects on wet deposition, and on the source-receptor relationships estimated from such modeling. The spring intensive period had much less precipitation than the summer case, and the WRF model simulated it relatively well.

The regional flow prediction can be influenced by the reanalysis data used to initialize and grid-nudge the WRF simulations. Larger differences were observed in the WRF results based on different reanalysis data in the summer campaign than in the spring campaign. For the 2010 summer period, the simulation using NARR data, commonly used for initializing the WRF model over North America domains, showed larger bias in comparison to the observations than the cases using other reanalysis data. Even with observational nudging, the fine-resolution domain still inherited differing features of the reanalysis data, which resulted in generating different regional wind patterns. The wind analysis at Grand Bay showed that the NARR case generated too much south-southwesterly wind compared with the other cases, and the observations actually showed north-northwesterly dominant winds during the campaign period. Similar wind patterns were generated among different sensitivity cases, but back-trajectory analyses were used to illustrate how even relatively small differences in regional wind fields can influence the modeled source-receptor relationships. In the example backward trajectory analysis of a summer 2010 mercury episode, the GFS-based simulation showed the air coming from the west, potentially bringing pollutants from emission sources to Grand Bay, while the NARR-based simulation showed air masses coming from the "clean" Gulf (i.e., with no large sources of mercury). The example trajectory analysis shown for the spring 2011 intensive also showed differences between the NARR- and GFS-based meteorological model results, but these were not as large as the difference shown in the summer 2010 episode. More differences between model results and measurements were observed between the various reanalysis datasets (used for model IC/BC/nudging) for the summer period than the spring campaign. According to the comparison of precipitation, the summer campaign period was stormier than the spring period. This may have led to higher uncertainties in the numerical model prediction during the summer period.

The 4-km grid spacing used here is generally considered to be at the borderline between the applicability of sub-grid cloud parametrizations (at larger grid spacing) and explicit approaches to generate convective updrafts (at smaller grid spacing). Future research will include simulations by using different microphysics schemes and cumulus parametrizations in the WRF model, as well as different grid sizes, to find the configuration(s) that give the best results. In addition, to improve the regional flow prediction, we will include the Grand Bay observations and SCAN dataset into the observational nudging, which may further reduce wind errors at those locations.

Acknowledgments

The authors thank Paul Kelly for his significant contribution to the data collection at the Grand Bay station.

Author Contributions

Fong Ngan wrote the majority of the manuscript and performed WRF simulations and model evaluations. Mark Cohen and Roland Draxler contributed valuable scientific insight and editing. Winston Luke and Xinrong Ren did the field campaigns at Grand Bay site, worked on data processing and quality control.

Conflicts of Interest

The authors declare no conflict of interest.

References

1. Driscoll, C.T.; Manson, R.P.; Chan, H.M.; Jacob, D.J.; Pirron, N. Mercury as a global pollutant: Sources, pathways, and effects. *Environ. Sci. Technol.* **2013**, *47*, 4967–4983.
2. Cohen, M.; Artz, R.; Draxler, R.; Miller, P.; Poissant, L.; Niemi, D.; Ratte, D.; Deslauriers, M.; Duval, R.; Laurin, R.; *et al.* Modeling the atmospheric transport and deposition of mercury to the Great Lakes. *Environ. Res.* **2004**, *47*, 4967–4983.
3. Cohen, M.; Artz, R.; Draxle, R. *Report to Congress: Mercury Contamination in the Great Lakes*; NOAA Air Resources Laboratory: Silver Spring, MD, USA, 2007.
4. Evers, D.C.; Wiener, J.G.; Basu, N.; Bodaly, R.A.; Morrison, H.A.; Williams, K.A. Mercury in the Great Lakes region: Bioaccumulation, spatiotemporal patterns, ecological risks, and policy. *Ecotoxicology* **2011**, *20*, 1487–1499.
5. Harris, R.C.; Pollman, C.; Landing, W.; Evans, D.; Axelrad, D.; Hutchinson, D.; Morey, S.L.; Rumbold, D.; Dukhovskoy, D.; Adams, D.H.; *et al.* Mercury in the Gulf of Mexico: Sources to receptors. *Environ. Res.* **2012**, *119*, 42–52.

6. Butler, T.J.; Cohen, M.D.; Vermeylen, F.M.; Likens, G.E.; Schmeltz, D.; Artz, R.S. Regional precipitation mercury trends in the eastern USA, 1998–2005: Declines in the Northeast and Midwest, no trend in the Southeast. *Atmos. Environ.* **2008**, *42*, 1582–1592.

7. Engle, M.A.; Tate, M.T.; Krabbenhoft, D.P.; Kolker, A.; Olson, M.L.; Edgerton, E.S.; DeWild, J.F.; McPherson, A.K. Characterization and cycling of atmospheric mercury along the central U.S. Gulf Coast. *Appl. Geochem.* **2008**, *23*, 419–437.

8. Ren, X.; Luke, W.T.; Kelley, P.; Cohen, M.; Ngan, F.; Artz, R.; Walker, J.; Brooks, S.; Moore, C.; Swartzendruber, P.; *et al.* Mercury speciation at a coastal site in the northern Gulf of Mexico: Results from the Grand Bay Intensive Studies in summer 2010 and spring 2011. *Atmosphere* **2014**, *5*, 230–251.

9. Nair, U.S.; Wu, Y.; Holmes, C.D.; Schure, A.T.; Kallos, G.; Walters, J.T. Cloud-resolving simulations of mercury scavenging and deposition in thunderstorms. *Atmos. Chem. Phys.* **2013**, *13*, 10143–10210.

10. Draxler, R.R. An overview of the HYSPLIT_4 modeling system for trajectories, dispersion and deposition. *Aust. Meteorol. Mag.* **1988**, *5*, 230–251.

11. Han, Y.J.; Holsen, T.M.; Hopke, P.K.; Yi, S.M. Comparison between back-trajectory based modeling and Lagrangian backward dispersion modeling for locating sources of reactive gaseous mercury. *Environ. Sci. Technol.* **2005**, *39*, 1715–1723.

12. Rolison, J.M.; Landing, W.M.; Luke, W.; Cohen, M.; Salters, V.J.M. Isotopic composition of species-specific atmospheric Hg in a coastal environment. *Chem. Geol.* **2013**, *336*, 37–49.

13. Gratz, L.E.; Keeler, G.J.; Marsik, F.J.; Barres, J.A.; Dvonch, J.T. Atmospheric transport of speciated mercury across southern Lake Michigan: Influence from emission sources in the Chicago/Gary urban area. *Sci. Total Environ.* **2013**, *448*, 84–95.

14. Lei, H.; Liang, X.Z.; Wuebbles, D.J.; Tao, Z. Model analyses of atmospheric mercury: Present air quality and effects of transpacific transport on the United States. *Atmos. Chem. Phys.* **2013**, *13*, 10807–10825.

15. Skamarock, W.C.; Klemp, J.B.; Dudhia, J.; Gill, D.O.; Barker, D.M.; Duda, M.G.; Huang, X.Y.; Wang, W.; Powers, J.G. A description of the advanced research WRF Version 3. In *NCAR Technical Note*; NCAR: Boulder, CO, USA, 2008; TN-475+STR.

16. W., W.; Lynch, A.H.; Rivers, A. Estimating the uncertainty in a regional climate model related to initial and lateral boundary conditions. *J. Appl. Climatol.* **2005**, *18*, 917–933.

17. Ngan, F.; Byun, D.W.; Kim, H.C.; Lee, D.G.; Rappenglueck, B.; Pour-Biazar, A. Performance assessment of retrospective meteorological inputs for use in air quality modeling during TexAQS 2006. *Atmos. Environ.* **2012**, *54*, 86–96.

18. Deng, A.; Stauffer, D.; Gaudet, B.; Dudhia, J.; Hacker, J.; Bruyere, C.; Wu, W.; Vandenberghe, F.; Liu, Y.; Bourgeois, A. Update on WRF-ARW end-to-end multi-scale FDDA system. In Proceedings of the 10th WRF Users' Workshop, Boulder, CO, USA, 23–26 June 2009; NCAR: Boulder, CO, USA, 2009.

19. Otte, T. The impact of nudging in the meteorological model for retrospective air quality simulations. Part I: Evaluation against national observation networks. *J. Appl. Meteor. Climatol.* **2008**, *47*, 1853–1867.

20. Lo, J.C.F.; Yang, Z.L.; Sr, R.A.P. Assessment of three dynamical climate downscaling methods using the Weather Research and Forecasting (WRF) model. *J. Geophys. Res.* **2008**, doi:10.1029/2007JD009216.

21. Godowitch, J.M.; Gilliam, R.C.; Rao, S.T. Diagnostic evaluation of ozone production and horizontal transport in a regional photochemical air quality modeling system. *Atmos. Environ.* **2011**, *45*, 3977–3987.

22. Rogers, R.; Deng, A.; Stauffer, D.R.; Gaudet, B.J.; Jia, Y.; Soong, S.; Tanrikulu, S. Application of the weatherresearch and forecasting model for air quality modeling in the San Francisco bay area. *J. Appl. Meteor. Climatol.* **2013**, *52*, 1953–1973.

23. Hegarty, J.; Coauthors. Evaluation of Lagrangian particle dispersion models with measurements from controlled tracer releases. *J. Appl. Meteor. Climatol.* **2013**, *52*, 2623–2637.

24. Gilliam, R.C.; Godowitch, J.M.; Rao, S.T. Improving the horizontal transport in the lower troposphere with four dimensional data assimilation. *Atmos. Environ.* **2012**, *53*, 186–201.

25. Mlawer, E.; Taubman, S.; Brown, P.D.; Iacono, M.; Clough, S. Radiative transfer for inhomogeneous atmosphere: RTTM, a validated correlated-k model for the longwave. *J. Geophys. Res.* **1997**, *102*, 16663–16682.

26. Dudhia, J. Numerical study of convection observed during the winter monsoon experiment using a mesoscale two-dimensional model. *J. Atmos. Sci.* **1989**, *46*, 3077–3107.

27. Xiu, A.; Pleim, J.E. Development of a land surface model. Part I: Application in a mesoscale meteorological model. *J. Appl. Meteor.* **2001**, *40*, 192–209.

28. Pleim, J.E.; Xiu, A. Development of a land surface model. Part II: Data assimilation. *J. Appl. Meteor.* **2003**, *42*, 1811–1822.

29. Pliem, J.E. A combined local and nonlocal closure model for the atmospheric boundary layer. Part I: Model description and testing. *J. Appl. Meteor. Climatol.* **2007**, *46*, 1383–1395.

30. Grell, A.G.; Devenyi, D. A generalized approach to parameterizing convection combining ensemble and data assimilation techniques. *J. Geophys. Res. Lett.* **2002**, doi:10.1029/2002GL015311.

31. Mesinger, F.; DiMego, G.; Kalnay, E.; Mitchell, K.; Shafran, P.C.; Ebisuzaki, W.; Jovic, D.; Woollen, J.; Rogers, E.; Berbery, E.H.; *et al.* North American regional reanalysis. *Bull. Am. Meteor. Soc.* **2006**, *87*, 343–340.

32. Kanamitsu, M. Description of the NMC global data assimilation and forecast system. *Weather Forecast.* **1989**, *4*, 334–342.

33. Kalnay, E.; Kanamitsu, M.; Kistler, R.; Collins, W.; Deaven, D.; Gandin, L.; Iredell, M.; Saha, S.; White, G.; Woollen, J.; *et al.* The NCEP/NCAR 40-year reanalysis project. *Bull. Am. Meteor. Soc.* **1996**, *77*, 437–471.

34. Saha, S.; Moorthi, S.; Pan, H.-L.; Wu, X.; Wang, J.; Nadiga, S.; Tripp, P.; Kistler, R.; Woollen, J.; Behringer, D.; *et al.* The NCEP climate forecast system reanalysis. *Bull. Am. Meteor. Soc.* **2010**, *91*, 1015–1057.

35. Chai, T.; Draxler, R.R. Root mean square error (RMSE) or mean absolute error (MAE)?—Arguments against avoiding RMASE in the literature. *Geosci. Model Dev.* **2014**, *7*, 1247–1250.

36. Gilliam, R.; Hogrefe, C.; Rao, S. New methods for evaluating meteorological models used in air quality applications. *Atmos. Environ.* **2006**, *40*, 5073–5086.

37. Yu, S.; Rohit, M.; Pleim, J.; Pouliot, G.; Wong, D.; Eder, B.; Schere, K.; Gilliam, R.; Rao, S. Comparative evaluation of the impact of WRF-NMM and WRF-ARW meteorology on CMAQ simulations for O3 and related species during the 2006 TexAQS/GoMACCS campaign. *Atmos. Pollut. Res.* **2012**, *3*, 149–162.

38. Wilks, D. *Statistical Methods in the Atmospheric Sciences*; Elsevier: Burlington, MA, USA, 2006; p. 549.

39. Lee, S.-H.; Kim, S.-W.; Angevine, W.M.; Bianco, L.; McKeen, S.A.; Senff, C.J.; Trainer, M.; Tucker, S.C.; Zamora, R.J. Evaluation of urban surface pa- rameterizations in the WRF model using measurements during the Texas Air Quality Study 2006 field campaign. *Atmos. Chem. Phys.* **2011**, *11*, 2127–2143.

40. Ngan, F.; Kim, H.; Lee, P.; Al-Wali, K.; Dornblaser, B. A study of nocturnal surface wind speed overprediction by the WRF-ARW model in southeastern Texas. *J. Appl. Meteor. Climatol.* **2013**, *52*, 2638–2653.

41. Chen, B.; Stein, A.F.; Castell, N.; de laRosa, J.D.; de laCampa, A.; Gonzalez-Castanedo, Y.; Draxler, R.R. Modeling and surface observations of arsenic dispersion from a large Cu-smelter in southwestern Europe. *Atmos. Environ.* **2012**, *49*, 114–122.

42. Zhang, D.; Zheng, W. Diurnal cycles of surface winds and temperatures as simulated by five boundary layer parameterizations. *J. Appl. Meteor.* **2004**, *1*, 157–169.

43. Tong, D.; Lee, P.; Ngan, F.; Pan, L. Investigation of Surface Layer Parameterization of the WRF Model and Its Impact on the Observed Nocturnal Wind Speed Bias: Period of Investigation Focuses on the Second Texas Air Quality Study (TexAQS II) in 2006. Available online: http://aqrp.ceer.utexas.edu/index.cfm (accessed on 16 February 2015).

2

Atmospheric Deposition History of Trace Metals and Metalloids for the Last 200 Years Recorded by Three Peat Cores in Great Hinggan Mountain, Northeast China

Kunshan Bao [1,*], Ji Shen [1], Guoping Wang [2] and Gaël Le Roux [3,4]

[1] State Key Laboratory of Lake Science and Environment, Nanjing Institute of Geography and Limnology, Chinese Academy of Sciences, Nanjing 210008, China;
E-Mail: jishen@niglas.ac.cn

[2] Key Laboratory of Wetland Ecology and Environment, Northeast Institute of Geography and Agroecology, Chinese Academy of Sciences, Changchun 130102, China;
E-Mail: wangguoping@neigae.ac.cn

[3] Université de Toulouse, INP, UPS, EcoLab (Laboratoire Ecologie Fonctionnelle et Environnement), ENSAT, Avenue de l'Agrobiopole, 31326 Castanet Tolosan, France;
E-Mail: gael.leroux@ensat.fr

[4] CNRS, EcoLab, 31326 Castanet Tolosan, France

* Author to whom correspondence should be addressed; E-Mail: ksbao@niglas.ac.cn

Academic Editor: Robert W. Talbot

Abstract: A large number of studies on trace metals and metalloids (TMs) accumulations in peatlands have been reported in Europe and North America. Comparatively little information is available on peat chronological records of atmospheric TMs flux in China. Therefore, the objective of our study was to determine the concentrations and accumulation rates (ARs) of TMs in Motianling peatland from Great Hinggan Mountain, northeast China, and to assess these in relation to establish a historical profile of atmospheric metal emissions from anthropogenic sources. To meet these aims we analyzed 14 TMs (As, Ba, Cd, Co, Cr, Cu, Mo, Ni, Pb, Sr, Sb, Tl, and Zn) and Pb isotopes (^{206}Pb, ^{207}Pb, ^{208}Pb) using ICP-AES and ICP-MS, respectively, in three peat sections dated by ^{210}Pb and ^{137}Cs techniques (approximately spanning the last 200 years). There is a general agreement in the elemental concentration profiles which suggests that all investigated elements were conserved in the Motianling bog. Three principal components were discriminated by principal component analysis (PCA) based on Eigen-values >1 and explaining 85% of the total variance of element concentrations: the first component representing Ba, Co, Cr, Mo, Ni, Sr and Tl reflected the lithogenic source; the second component covering As, Cu and Sb, and Cd is associated with an anthropogenic source from ore mining and processing; the third component (Pb isotope, Pb and Zn) is affected by anthropogenic Pb pollution from industrial manufacturing and fossil-fuel combustion. The pre-industrial background of typical pollution elements was estimated as the average concentrations of TMs in peat samples prior to 1830 AD and with a $^{207}Pb/^{206}Pb$ ratio close to 1.9. ARs and enrichment factors (EFs) of TMs suggested enhanced metal concentrations near the surface of the peatland (in peat layers dated from the 1980s) linked to an increasing trend since the 2000s. This pollution pattern is also fingerprinted by the Pb isotopic composition, even after the ban of leaded gasoline use in China. Emissions from coal and leaded gasoline combustions in northern China are regarded as one of the major sources of anthropogenic Pb input in this region; meanwhile, the long-distance transportation of Pb-bearing aerosols from Mongolia should be also taken into consideration. The reconstructed history of TMs' pollution over the past *ca.* 200 years is in agreement with the industrial development in China and clearly illustrates the influence of human activities on local rural environments. This study shows the utility of taking multi-cores to show the heterogeneity in peat accumulation and applying PCA, EF and Pb isotope methods in multi-proxies analyses for establishing a high resolution geochemical metal record from peatland.

Keywords: peatlands; trace metals; environmental pollution; historical trends; ^{210}Pb-dating; lead isotopes; Asia

1. Introduction

Trace metals and metalloids (referred to here as TMs) have been emitted by human activities and dispersed into the environment since the beginning of metallurgy [1]. They usually have many adverse impacts on environmental and human health because of their toxic, persistent and bio-accumulative nature. In addition to the natural sources including volcanic activity, soil erosion, and biologically driven reduction processes in the ocean, anthropogenic sources are the major contributions to abundance of TMs in the environment and principally include fossil-fuel combustion, industrial manufacturing, ore mining and processing, and incineration of urban, medical and industrial wastes [2]. Human activities have been shown to increase local, regional and global fluxes of TMs to the atmosphere, and anthropogenic TMs' fluxes clearly exceed the prehistoric levels of atmospheric deposition. As a result, governments around the world have negotiated a global legally binding instrument to control TM emissions from human activities, such as the Protocol on Long-range Transboundary Air Pollution of Heavy Metals for the United Nation (UN)/Economic Commission for Europe (ECE) [3]. A considerable percentage of TMs released to the atmosphere is often carried by long range transport, deposited by precipitation or as aerosols, and stored in and between aquatic and terrestrial ecosystems in remote areas. Long term records to quantify accumulation of TMs in natural geological archives are helpful to differentiate and document the temporal trends of natural *versus* anthropogenic TMs [4–8]. The magnitude and history of changes in past metal deposition have been studied in a variety of environmental records in the Northern Hemisphere, such as ice core [9], lake sediment [10,11] and ombrotrophic peat [12,13].

Ombrotrophic peatlands are hydraulically isolated, and receive all their chemical constituents via direct deposition from atmosphere or via uptake from the atmosphere by vegetation [14]. As a result, records from ombrotrophic peatlands can provide valuable information about atmospheric inputs of TMs [15,16]. Assuming that most metals deposited on the surface of ombrotrophic peatlands are sequestered by living or dead organic matter and are immobile, they can also provide a reliable indication of the changing deposition and, thus, of the content of elements in the air [17,18].There have been a large number of studies in the past three decades on the use of peat bogs as archives of atmospheric TMs' deposition and most of them have been carried out in Europe and North America [19–23] and references therein. In contrast to those extensive amounts of work, few similar studies have been conducted in China. Most of the latter focused on either on Hg [24,25] or Pb [26,27]. Another two emerging studies used inorganic geochemistry of the peat record to reconstruct the history of atmospheric dust fluxes, and focused on major lithogenic elements such as Al, Ca, Fe, Mg and Ti and other trace elements such as rare earth elements (REE), Sc, Y and Th [28,29].

Such investigations above have focused on geochemical analysis of peat sampled at different depths and establishment of rates of peat accumulation through ^{14}C or ^{210}Pb dating to provide information about pollution chronologies. The fluxes of atmospheric TMs deposition could be calculated through combining the rate of peat accumulation, dry bulk density and TMs concentrations. This approach assumed that TMs are immobile and well preserved in peatlands. However, a peatland is an active system as a result of both accumulation and decomposition processes of organic matter. The mobility of TMs, especially in the peat surface layer (*i.e.*, the acrotelm) is linked to various factors such as adsorption on oxyhydroxides [30], variation in pH [31], decomposition of organic matter [32], and uptake and recycling by plants [33], and thus their behavior is not always well understood, despite the numerous studies undertaken. Lead is by far the most intensively studied heavy metal in peat and is considered by most authors as an immobile element in peat [12,34]. Moreover, the isotopic composition of Pb is regarded to be a powerful tool for distinguishing natural from anthropogenic sources [35] and could also provide a support for Pb immobility by comparing Pb isotopic composition in dated herbarium Sphagnum samples and in ^{210}Pb-dated peat samples from the same age and same location [4,36]. Zinc usually accompanies Pb in geochemical studies of peat and both of them are often associated in some important sources of pollution including sulfide minerals and coals [21].However, to some extent, Zn is relatively mobile due to bioaccumulation, water table fluctuations and pH conditions [37,38], so the use of Zn as a tracer is still under debate. Similar to Zn, the behavior of Cu is demonstrated to be either immobile [39,40] or mobile [18,41]. Compared to Pb, Zn and Cu, less studies for Ni behavior in peat and contradictory conclusions existed from different researches. In a study of atmospheric deposition in peat since the late glacial period, thus focused on the catotelm, Krachler *et al.* [42] found that Ni is mainly immobile; whereas, Ukonmaanaho *et al.* [43] argued that Ni is mobile and affected by leaching and pH condition in the acrotelm. With respect to the other TMs like As, Cd, Cr, Co, Sb and V, there is an increasing concern of their behaviors in ombrotrophic peatlands; their sources are mainly anthropogenic emissions and their distributions in peat are usually comparable to the Pb profile [44–46].

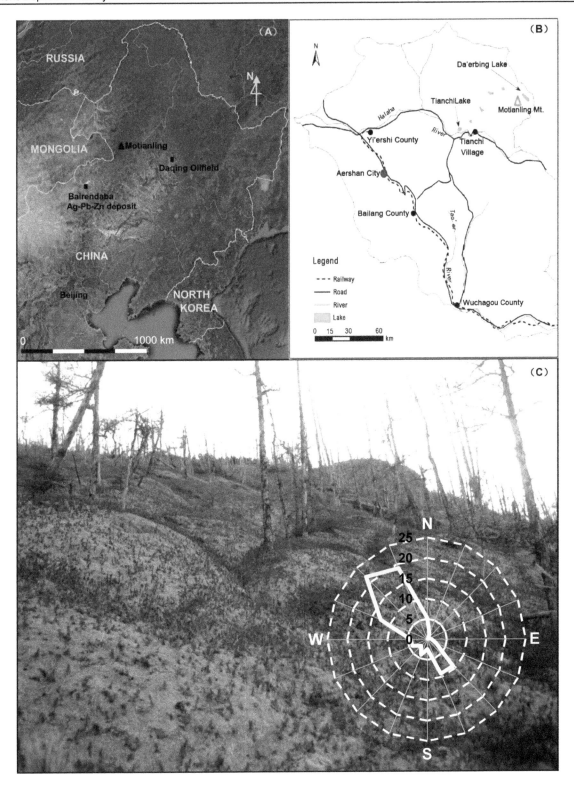

Figure 1. (A) Regional map showing the location of Aershan in northeast Asia; **(B)** Local map of Aershan city showing the sampling site of Motianling peatland in Aershan city in Great Hinggan Mountains, NE China; **(C)** Photo showing the early summer scene in the studied peatland (photo in June 2009). Average wind frequency (2005–2010) is inserted in **(C)**, showing the prevailing wind in Aershan is mainly from the northwest and southeast (%).

When a large geochemical dataset is generated in such a multi-tracer study of peat records, principal component analysis (PCA) is often applied to reduce the number of observed variables to a smaller number of factors and to identify sources and processes related to the distribution of multiple elements [47–50]. This statistical technique is used to a single set of variables (*i.e.*, geochemical elements) to discern which set of variables form coherent subsets that are relatively independent of one another. Those correlated variables are combined into factors which should be representative of the underlying process. Loadings measure the correlations among these variables and relate to the specific association between the factors and original variables [51]. Another common approach to distinguish between natural and anthropogenic sources for elements in the environment is to calculate a normalized enrichment factor (EF) for metal concentrations above uncontaminated background levels [52–54]. The calculation of EF is usually based on a conservative (lithogenic) element indicative of mineral matter, such as Ti; in addition to the reference element, such a calculation is often relative to a reference material, either external (*i.e.*, the continental crust) or internal (*i.e.*, the mean of several low concentration samples selected from the deeper, pre-industrial levels of cores) [54]. The interpretation of EF is that an element with an EF value near unity has a probable source in crustal material and elements with EF value a few to many times larger than unity could be mainly industrial origin [55].

China has been one of the fast-growing economics of the late 20th and early 21st centuries, and heavy metal emissions have been increased in parallel with rapid industrialization and urbanization which has led to elevated metal concentrations in air, water and soil [27]. It has become a growing concern for human health; however, the temporal variation of atmospheric metal pollution is still not well understood due to the lack of long-term environmental records of atmospheric metal deposition. There are extensive peatlands developed on the alpine areas in northeast China, such as Great Hinggan Mountains [56]. Recently, we did an initial study of geochemical records through ^{210}Pb-dated cores from a peatland in this region and provided a preliminary estimate of atmospheric Pb and soil dust fluxes over the past 200 years [26,28]. However, a complete representation of the factors affecting TMs accumulation in peatlands is still absent in China. Therefore, this study presents a synthesis on TMs in the same peatland in Motianling of Great Hinggan Mountain to give an overview of present and past contamination in northeast China (Figure 1). The objectives are to determine the concentrations, accumulation rates and inventories of As, Ba, Cd, Co, Cr, Cu, Mo, Ni, Pb, Sr, Sb, Tl, and Zn in Motianling peatland using a peat multi coring approach, and to assess TMs sources through the use of stable Pb isotopes (^{206}Pb, ^{207}Pb, ^{208}Pb), calculation of EF and PCA analyses.

2. Material and Methods

2.1. Brief Description of Previous Studies

A detailed description of three peat cores, the procedures used to section it and to prepare the peat samples, as well as the methods used to measure the basic physical and chemical indicators (Water content, dry bulk density, ash content, *etc.*) and major and trace elements using ICP-AES (Al, Ca, Fe, Mn, V, Ti and Pb), as well as the age dating undertaken using ^{210}Pb and ^{137}Cs, is provided elsewhere [26,28]. The following is a brief summary of these procedures: at the Motianling peat deposit on the north facing slope (*ca.* 30°) of Motianling mountain in northeast China (Figure 1), MP1 and MP2 columns were collected in October of 2008 and MP3 column was collected in June of 2009 by digging a pit. On the straight face of the pit, we first isolated a 15 cm × 15 cm column of peat by removing material on either side, and then carefully cut the 1 or 2 cm thick layers from this column and placed in plastic bags. Individual slices were sub-sampled, prepared, and measured for water content (WAT, %), dry bulk density (DBD, g·cm^{-3}), weight loss on ignition (LOI, %), ash content (ASH, %) [26]. Selected elements (Al, Ca, Fe, K, Mn, Mg, Na, Pb, Ti, V) were determined by ICP-AES (Shimadzu Corporation, Japan). The detailed process of sample preparation and results of Al, Ca, Fe, Mn, V and Ti were given in Bao *et al.* [28], and Pb data of MP1 and MP2 were reported in Bao *et al.* [26]. Dried bulk peat samples were age dated using ^{210}Pb and ^{137}Cs techniques through a low-background γ-ray spectrometer with a high pure Ge semiconductor (ORTEC Instruments Ltd., Oak Ridge, TN, USA) as described previously [26]. The unsupported ^{210}Pb (excess ^{210}Pb) activities were gradually declined with depth downward in all cores, and they became negligible at depths of 58 cm for MP1, 64 cm for MP2 and 60 cm for MP3, respectively.

2.2. Peat pH, Grain-Size, TOC and TN

The pH of peat was not measured when the fresh samples were collected and thus determined from our frozen samples which were thawed at room temperature. Water could be squeezed out of most of samples gently by hand, but some without water due to the lower water content at the bottom of core. The squeezed water or 1 g bulk peat was added into 5 mL deionized water. They were stirred adequately (*ca.* 30 min) and measured by a pH meter after 30 min standing. Total organic carbon (TOC) and total nitrogen (TN) were determined using the FlashEA1112 elemental analyzer (Thermo Finnigan, Milan, Italy). The quality of TOC and TN measurement was checked by periodic analysis of a certified reference material (GSS-1, Brown soil), with an error <1%. Granulometry of the samples was determined through the measurement of the grain size of peat ash prepared by ignition of dry samples in a muffle furnace at 700°C ± 50°C for 4 h [57]. Ash sample (*ca.* 0.2 g) was dissolved in 10 mL distilled water with the addition of sodium hexametaphosphate (($NaPO_3$)$_6$, 3.6%) as a dispersant. Sample solutions were ultrasonicated for 10 min to facilitate dispersion for a concentration of 10%–20% prior to analysis by Mastersizer 2000 Laser Particle Size Analyser (Malvern Ltd., Worcestershire, UK) with a measurement range of 0.02–2000 μm. The repeated measurement error is generally less than 3%.

2.3. Determination of Trace Elements

Selected TMs (As, Ba, Cd, Co, Cr, Cu, Mo, Ni, Pb, Sr, Sb, Tl and Zn) and stable Pb isotopes (^{206}Pb, ^{207}Pb, ^{208}Pb) were measured by a Quadrupole Inductively Coupled Plasma Mass Spectrometry (Q-ICP-MS 7700x, Agilent Technologies, Santa Clara, CA, USA) at the State Key Laboratory of Lake Science and Environment, Nanjing Institute of Geography and Limnology, CAS. Because total concentrations of Pb were measured for MP1 and MP2 cores [26], here we just measured Pb contents for MP3 core which were combined with the previous results. Sample aliquots were dried at 105 °C for 12 h and ground using an agate mortar. The fine sample was weighted accurately (0.2000–0.2059 g) into a TFM-PTFE in-liner vessel ("bomb") with the stainless steel pressure digestion system (DAB-2, Berghof, Berchtesgaden, Germany). 2.5 mL HNO_3 (high purity obtained by sub-boiling distillation of the analytical-grade reagent in an I.R. distiller (BSB-939, Berghof, Berchtesgaden, Germany)) and 0.5 mL H_2O_2 (Baker ACS Reagent) were added and the PTFE vessels were close heated at 200°C–220 °C for 3 h. After cooling, 1 mL HF (Baker ACS Reagent) and 0.5 mL $HClO_4$ (Fisher Trace Metal Grade) were added and they were heated again until the white smoke from the in-liner disappeared. After cooling again, 0.5 mL HNO_3 (high purity) and 2.5 mL deionized water (>18 MΩ·cm) were added and they were heated at 150°C–180°C for 5 min. Last, they were cooled to ambient temperature for a minimum of 2 h, and the digested samples were centrifuged and the supernatant was transferred to a 50 mL centrifuge tube along with the washings from the TFM-PTFE in-liner. The sample solutions were brought up to a final volume of 25 mL with deionized water for TM analysis. The above solutions were diluted to a concentration of 8–10 $\mu g \cdot L^{-1}$ Pb for measuring the isotopic composition of Pb [35]. NIST SRM 981 was used as an international standard reference material for every five samples to ensure the precision of the measurement process. The relative standard deviations (RSD) for $^{206}Pb/^{207}Pb$ and $^{208}Pb/^{206}Pb$ were <0.12% and <0.07%, respectively. The accuracy of the measurements was tested on replicate analyses of some randomly selected samples and the results of TMs concentrations and Pb isotope ratios were summarized in Table 1 and expressed as mean ± standard deviation.

2.4. Calculations of EFs and ARs

Two criterions were used to establish a local "baseline" concentration for each core by taking the mean of several low concentration samples selected from (1) the deep, least impacted level of peats prior to the industrial revolution (i.e., 1830); (2) the peats with the $^{206}Pb/^{207}Pb$ ratio larger than 1.19 (typically radiogenic value). The determined background values of the lithogenic element (Ti) and the six pollution elements (As, Cd, Cu, Pb, Sb and Zn) identified by PCA were summarized in Table 2. They are comparable to the concentrations in standard earth materials such as average shale [58] or average crustal values [59,60].The difference between the total concentrations of TMs (TM_{total}) and the background concentrations of TMs ($TM_{background}$) provided an estimate of excess TM concentration (TM_{ex}) from human activities (Equation (1)).The EFs relative to the local background content of TMs were calculated according to Equation (2) detailed in a previous study [13]. The annual AR TMs were calculated from Equation (3) proposed by Givelet et al. [61].

$$TM_{ex} = TM_{total} - TM_{background} \tag{1}$$

$$EF_{local} = (TM_{total}/Ti_{total})_{sample}/(TM_{total}/Ti_{total})_{background} \tag{2}$$

$$AR\ TM\ (mg \cdot m^{-2} \cdot yr^{-1}) = TM_{total}\ (\mu g \cdot g^{-1}) \times DBD\ (g \cdot cm^{-3}) \times SR\ (cm \cdot yr^{-1}) \times 10 \tag{3}$$

where $(TM_{total})_{sample}$ is the total concentration of the elements measured in the peat sample; $TM_{background}$ is the abundance measured at the bottom section of the bog, by average the value of samples with $^{206}Pb/^{207}Pb \geq 1.9$ and taken from pre-1830 layers, representing the natural pre-industrial elemental background; Ti is used as the conservative reference element; DBD is the bulk density of the peat and SR is the sedimentation rate of the peat. The methodology to calculate sedimentation rate (SR, $cm \cdot yr^{-1}$) and peat accumulation rate (PAR, $g \cdot cm^{-2} \cdot yr^{-1}$) are described in a previous report [26].

Table 1. Replicate analyses of trace metal and metalloid (TM) concentrations (mg·kg⁻¹) and Pb isotope ratios (^{208}Pb/^{206}Pb and ^{206}Pb/^{207}Pb) with respective to some randomly selected samples from Motianling peatland and the results are expressed as mean ± standard deviation.

Sample Number	Depth (cm)	As	Ba	Cd	Co	Cr	Cu	Mo (mg·kg⁻¹)	Ni	Pb *	Sr	Sb	Tl	Zn	^{208}Pb/^{206}Pb	^{206}Pb/^{207}Pb
MP1-1	0.5	3.45 ± 0.07	146.5 ± 2.12	0.35 ± 0.01	2 ± 0	11.4 ± 0.42	7.15 ± 0.07	0.75 ± 0.07	5.40 ± 0.14		61.15 ± 1.06	0.45 ± 0.01	0.14 ± 0	66.40 ± 1.70		
MP1-5	4.5	1.95 ± 0.07	81.85 ± 0.07	0.1 ± 0	1 ± 0	5.8 ± 0	5.4 ± 0.14	0.3 ± 0	3 ± 0		30 ± 0.14	0.26 ± 0.04	0.07 ± 0	58.85 ± 1.63	2.1134 ± 0.0.0025	1.1639 ± 0
MP1-10	9.5	1.8 ± 0	73.75 ± 0.07	0.16 ± 0.01	0.95 ± 0.07	6 ± 0.28	3.75 ± 0.07	0.3 ± 0	3.1 ± 0.14		28.75 ± 0.49	0.21 ± 0.02	0.06 ± 0	26.9 ± 0.85		
MP1-21	31	1.65 ± 0.07	110.5 ± 2.12	0.1 ± 0	1.4 ± 0	6.8 ± 0.28	5.5 ± 0	0.35 ± 0.07	4 ± 0.14		42.7 ± 0.57	0.23 ± 0.01	0.06 ± 0	33.6 ± 1.84		
MP1-22	33	1.55 ± 0.07	83.75 ± 2.62	0.12 ± 0	1.35 ± 0.07	5.05 ± 0.78	4.9 ± 0.28	0.25 ± 0.07	3.6 ± 0		35.5 ± 0.57	0.18 ± 0.01	0.04 ± 0	32.3 ± 2.83	2.1089 ± 0.0027	1.1677 ± 0.0037
MP1-31	51	8.35 ± 0.64	443.5 ± 3.54	0.17 ± 0	13.75 ± 0.21	66.8 ± 1.7	23.4 ± 0.14	1.55 ± 0.07	29.45 ± 0.49		153.5 ± 2.12	0.71 ± 0.04	0.4 ± 0	77.05 ± 1.91		
MP1-35	59	6.7 ± 0.28	415 ± 5.66	0.17 ± 0	15.05 ± 0.49	60.35 ± 0.78	24.65 ± 0.35	1.5 ± 0	30.5 ± 1.56		146 ± 2.83	0.57 ± 0.01	0.36 ± 0.01	54.8 ± 0.99	2.0784 ± 0.0005	1.1912 ± 0.0030
MP2-4	3.5	1.65 ± 0.07	35.05 ± 0.78	0.15 ± 0	0.9 ± 0	5 ± 0.28	3.75 ± 0.35	0.3 ± 0	2.7 ± 0		18.85 ± 0.07	0.22 ± 0	0.05 ± 0	33.6 ± 1.84		
MP2-13	13	2.15 ± 0.07	65.05 ± 0.92	0.14 ± 0	1.5 ± 0	7.6 ± 0.14	5.85 ± 0.07	0.3 ± 0	4.05 ± 0.07		33.7 ± 0.14	0.23 ± 0	0.09 ± 0	32 ± 1.27	2.1100 ± 0	1.1666 ± 0.0004
MP2-14	15	1.95 ± 0.07	71.85 ± 1.63	0.19 ± 0	1.8 ± 0	6.8 ± 0.42	4.7 ± 0	0.3 ± 0	4.1 ± 0.14		34.95 ± 0.78	0.25 ± 0.01	0.08 ± 0	29.25 ± 0.64		
MP2-25	37	1.75 ± 0.07	177 ± 1.41	0.1 ± 0	3.1 ± 0	8.9 ± 0	8.25 ± 0.07	0.4 ± 0	7.95 ± 0.07		86.1 ± 0.85	0.18 ± 0	0.08 ± 0.01	91.25 ± 32.17		
MP2-38	63	7.1 ± 0.14	432.5 ± 0.71	0.13 ± 0	9.15 ± 0.07	64.35 ± 0.64	23.15 ± 0.07	1.8 ± 0	24.5 ± 0.57		168 ± 0	0.68 ± 0	0.42 ± 0	78.85 ± 2.90	2.0715 ± 0.0007	1.1932 ± 0.0008
MP3-2	3	17.9 ± 0.99	993.5 ± 12.02	0.88 ± 0.11	18.55 ± 1.48	65.25 ± 7	103.95 ± 7.14	4.4 ± 0.14	50.75 ± 5.02	109.5 ± 6.36	414 ± 7.07	2.85 ± 0.35	0.99 ± 0.08	312.5 ± 31.82		
MP3-11	21	15.2 ± 1.13	1103.5 ± 14.85	0.09 ± 0.01	20.85 ± 0.21	65.1 ± 5.66	96.05 ± 2.19	3.85 ± 0.07	148.5 ± 34.65	45.2 ± 0.85	674.5 ± 6.36	1.7 ± 0.14	0.6 ± 0.01	197.5 ± 6.36		
MP3-22	43	12.9 ± 0.14	1358 ± 2.83	0.34 ± 0.02	24.95 ± 0.21	62.25 ± 2.47	101.7 ± 24.47	2.6 ± 0	128.4 ± 88.53	38.8 ± 3.54	665 ± 2.83	1.45 ± 0.35	0.56 ± 0	184 ± 0		
MP3-33	65	9.8 ± 0	859 ± 11.31	0.28 ± 0.03	15.05 ± 0.07	80.65 ± 2.62	54.5 ± 3.68	2.35 ± 0.07	45.3 ± 3.68	28.6 ± 0.85	377.5 ± 3.54	1.35 ± 0.07	0.63 ± 0.01	101.85 ± 5.87		
MP3-39	77	9.1 ± 0.14	644 ± 8.49	0.17 ± 0.04	10.5 ± 0.28	76.95 ± 8.41	35.25 ± 1.48	2.15 ± 0.07	32.7 ± 0.85	26.65 ± 0.35	213.5 ± 2.12	2.3 ± 0.14	0.65 ± 0.03	97.9 ± 1.7	2.0830 ± 0.0014	1.1887 ± 0.0010

* Indicates that samples of MP3 core were just measured for Pb concentration by ICP-MS; Pb content data of MP1 and MP2 were from Bao et al. [26] and not shown here.

Table 2. Background concentration values for lithogenic element (Ti) and pollution elements identified by PCA (As, Cd, Cu, Pb, Sb and Zn) for the three studied cores, based on the $^{206}Pb/^{207}Pb$ ratio larger than 1.19. Data is presented as follows: average $\pm$ standard deviation (SD). The numbers in the bracket following cores are numbers of samples.

Element	MP1 ($n = 7$)	MP2 ($n = 3$)	MP3 ($n = 3$)	Earth's Crust [59]	Continental Shale [58]	Continental Crust [60]
Ti (mg/g)	3.19 ± 1.29	4.61 ± 1.68	4.13 ± 0.81	4.5	5.7	4.01
As (mg/kg)	5.36 ± 1.32	2.58 ± 0.54	1.75 ± 0.15	1.7	1.8	1.7
Cd (mg/kg)	0.17 ± 0.00	0.14 ± 0.03	0.21 ± 0.02	0.13	0.2	0.1
Cu (mg/kg)	14.55 ± 3.87	8.26 ± 1.84	5.74 ± 0.99	47	55	25
Pb (mg/kg)	28.36 ± 7.53	43.80 ± 16.40	36.89 ± 5.20	16	20	14.8
Sb (mg/kg)	0.62 ± 0.08	0.26 ± 0.06	0.25 ± 0.17	0.5	0.2	0.3
Zn (mg/kg)	67.88 ± 14.42	70.91 ± 32.52	72.71 ± 19.06	83	70	65

2.5. Data Statistical Analysis

Values of mean, standard deviation, minimum and maximum were calculated for the variables of peat deposition, elemental concentrations and ARs. PCA analysis was applied to identify and interpret the underlying (latent) factors affecting chemical elements in peat record. Pearson correlation analysis was performed to examine relationships among the individual parameters. Before PCA and Pearson correlation analyses, all data were converted to Z-scores calculated as per Equation (4). Principal components were extracted applying a varimax rotation to the complete dataset combining the data of all cores. These procedures were performed using SPSS 11.5 software package [62]. Statistical significance was determined at the $P = 0.05$ level unless indicated otherwise.

$$Z_{\text{-score}} = (X_i - X_{avg})/X_{std} \qquad (4)$$

where X_i, X_{avg}, and X_{std} are the given value, the average and the standard deviation of a variable in a given sample, respectively.

3. Results

3.1. TOC, TN, Peat pH and Grain-Size Distribution

Distribution of TOC was characterized by high values up to 45% in the topmost section followed by a subsequent decrease at the deeper layers, and the minimum value was 9.5% at the bottom of core. For TN, there was an increasing trend with depth near the surface, with an obvious peak at *ca.* 40 cm (MP1), 55 cm (MP2) and 50 cm (MP3), and below these depths, variation of TN was similar to that of TOC. The C/N ratios were calculated to represent the degree of peat decomposition and had a steadily decreasing trend with depth except for the abnormal high value at the bottom of MP1 core (Figure 2). For MP3 core, the C/N ratio has a quite high value (up to 128) at *ca.* 20 cm depth. The pH values of the three cores were quite consistent and increased with depth downward gradually (Figure 2). The following grain-size fractions were determined: clay (<2 μm), fine silt (2–25 μm), coarse silt (25–53 μm), fine sand (53–250 μm), and coarse sand (>250 μm). The profiles variations of main fractions with depth in MP1, MP2 and MP3 cores are also shown in Figure 2. The fine silt composition accounted for 68.6% ± 8.3%

of MP1, 72.6% ± 8.5% of MP2 and 71.4% ± 9.3% of MP3, respectively. The grain-size distribution is nearly homogeneous, exclusively assembled with clay-silt fine grain fractions over the entire core length (92.8% ± 5.2% < 53 μm for MP1, 94.4% ± 4.9% < 53 μm for MP2, 94.8% ± 3.5% < 53 μm for MP3).

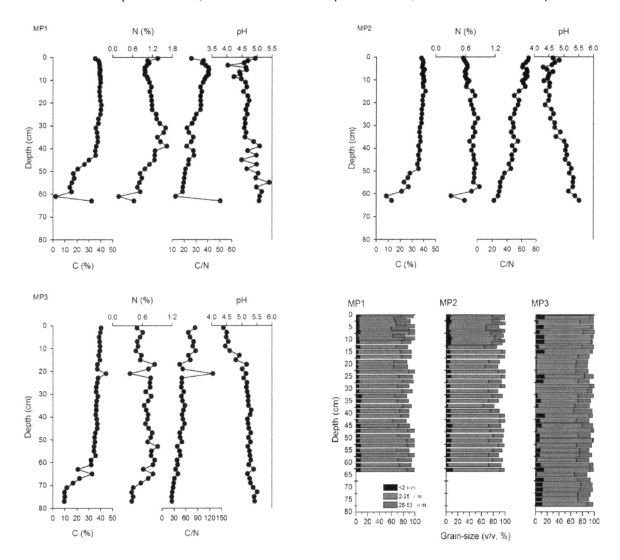

Figure 2. Depth profiles of total organic carbon content (TOC, %), total nitrogen content (TN, %) as well as their ratios (C/N), and grain-size distribution (v/v, %) of particles in the peat ash from Motianling peatland in Great Hinggan Mountains, northeast China.

3.2. Elemental Concentrations and Pb Isotope Ratios

Means of element concentrations (As, Ba, Cd, Co, Cr, Cu, Mo, Ni, Pb, Sb, Sr, Tl and Zn) in the different peat cores (Table 3) show that the variability among the three cores is generally low (<two-fold difference) except that there are slightly higher values of Sr, Tl and Cr in the MP3 core (*ca.* three-fold, 4.5-fold and 5.5-fold than other cores). Depth distributions of TMs are shown in Figure 3. They are all characterized by obvious elevated concentrations in the surface section as well as at the bottom section. All elements show smaller variation with a factor of lower than 10; however, great variability of concentrations was found for Cr, Co and Cu in MP1, Ba and Co in MP2, and Sb in MP3, with a factor of above 10 (Table 3). Lead isotopic composition involving $^{206}Pb/^{207}Pb$ and $^{208}Pb/^{206}Pb$ revealed quite

similar behavior for the three profiles (Figure 3). Ratios of ^{206}Pb/^{207}Pbare generally increasing with depth downward, with the minimum value of 1.156 at 9.5 cm depth (MP1), 1.157 at 2.5 cm depth (MP2), and 1.159 at 3 cm depth (MP3).

Table 3. Mean, maximum and minimum concentrations of elements (mg·kg^{-1}) in peat cores taken from three cores in Motianling peatland of the Great Hinggan Mountains, northeast China.

Element	Min	Max	Mean	Factor$_{max/min}$	Min	Max	Mean	Factor$_{max/min}$	Min	Max	Mean	Factor$_{max/min}$
	MP1 (n = 37)				MP2 (n = 38)				MP3 (n = 39)			
As	1.25	6.91	2.84	5.53	1.58	3.70	2.08	2.34	0.78	3.20	1.62	4.10
Ba	65.29	222.95	115.36	3.41	35.08	432.25	135.54	12.32	163.64	323.39	229.62	1.98
Cd	0.08	0.35	0.15	4.38	0.09	0.29	0.14	3.22	0.05	0.39	0.17	7.80
Cr	4.00	58.79	13.34	14.70	4.98	34.35	9.11	6.90	32.96	81.36	50.15	2.47
Co	0.96	11.26	2.99	11.73	0.92	9.14	2.82	9.93	3.33	11.70	5.24	3.51
Cu	2.76	29.60	8.99	10.72	4.92	10.03	7.13	2.04	3.58	10.32	5.25	2.88
Mo	0.20	1.33	0.54	6.65	0.28	0.96	0.40	3.43	0.60	1.34	0.92	2.23
Ni	2.74	20.25	6.87	7.39	2.71	20.51	6.75	7.57	8.88	21.51	13.88	2.42
Pb	14.69 *	110.40 *	47.59 *	7.52 *	29.49 *	261.98 *	78.01 *	8.88 *	20.10	123.00	44.51	6.12
Sb	0.16	0.70	0.33	4.38	0.17	0.41	0.23	2.41	0.05	0.50	0.21	10.00
Sr	27.64	134.27	55.06	4.86	18.86	167.92	63.68	8.90	126.24	236.44	167.28	1.87
Tl	0.03	0.20	0.09	6.67	0.05	0.22	0.09	4.40	0.27	0.69	0.43	2.56
Zn	26.95	266.42	57.09	9.89	25.70	210.32	65.61	8.18	57.81	188.53	89.65	3.26

* Indicates that Pb content data of MP1 and MP2 were from Bao *et al.* [26] and those of MP3 were measured by ICP-MS.

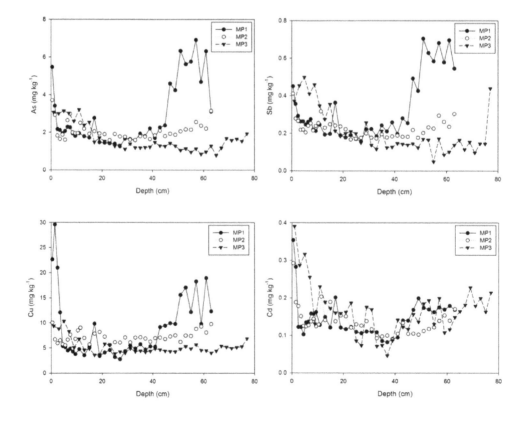

Figure 3. *Cont.*

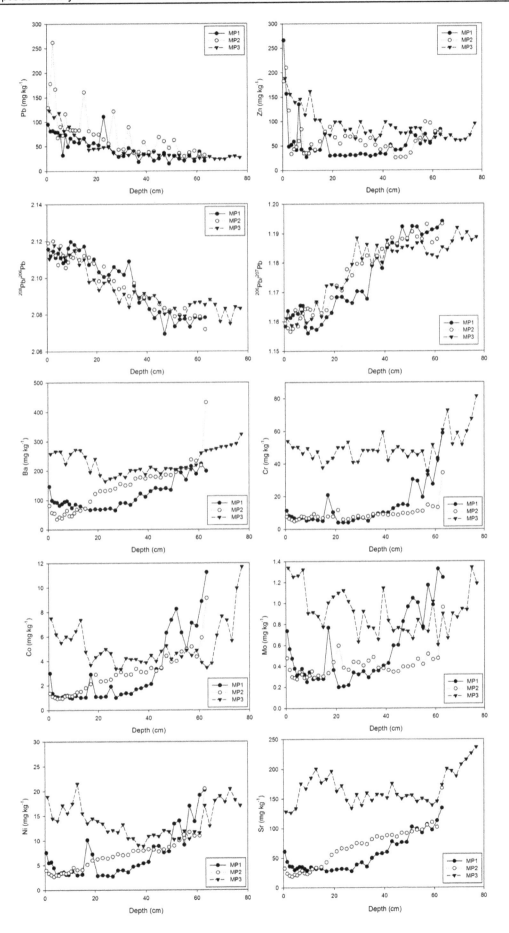

Figure 3. *Cont.*

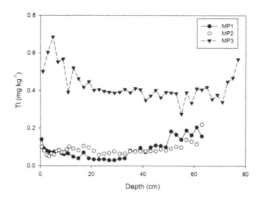

Figure 3. Concentration records of elements (As, Ba, Cd, Co, Cr, Cu, Mo, Ni, Pb, Sb, Sr, Tl and Zn; mg·kg^{-1}) and lead isotope ratios (^{206}Pb/^{207}Pb and ^{208}Pb/^{206}Pb) in the three cores sampled at Motianling peatland of Great Hinggan Mountains, northeast China. Pb content data of MP1 and MP2 were from Bao et al. [26].

Table 4. Varimax rotated factor matrices for the transformed (Z-scores) elemental concentrations in the first three principal components obtained by PCA analysis of the entire Motianling peatland geochemical dataset (MP1, MP2 and MP3). Significant (>0.4 and <−0.4) factor loadings are designated in bold. Proportion of the total data variance captured by a component is given as Eigenvalue, % Variance and Cumulative variance at the bottom of the table. Percentage of the variance of each element explained by each component and accounted for all the extracted principle components are given in four columns from the right.

	Factor Loadings			Partition of Communality (%)			Extraction Communalities (%)
	PC1	PC2	PC3	PC1	PC2	PC3	Total
Tl	**0.941**	−0.138	0.096	88.5	1.9	0.9	91.4
Cr	**0.941**	−0.097	−0.123	88.5	0.9	1.5	90.9
Sr	**0.932**	−0.147	−0.271	86.9	2.2	7.3	96.3
Ni	**0.899**	0.191	−0.264	80.8	3.6	6.9	91.4
Mo	**0.877**	0.310	−0.104	76.9	9.6	1.1	87.6
Ba	**0.859**	0.053	−0.352	73.8	0.3	12.4	86.5
Co	**0.777**	0.352	−0.394	60.4	12.4	15.5	88.3
As	−0.006	**0.957**	−0.048	0	91.6	0.2	91.9
Sb	0.086	**0.916**	0.032	0.7	83.9	0.1	84.8
Cu	−0.047	**0.865**	0.025	0.2	74.8	0.1	75.1
Cd	**0.455**	**0.535**	**0.516**	20.7	28.6	26.6	76.0
^{206}Pb/^{207}Pb	0.316	0.068	**−0.894**	10.0	0.5	79.9	90.3
^{208}Pb/^{206}Pb	−0.317	−0.096	**0.888**	10.0	0.9	78.9	89.8
Pb	−0.230	0.097	**0.789**	5.3	0.9	62.3	68.5
Zn	**0.512**	0.379	**0.561**	26.2	14.4	31.5	72.1
Eigenvalue	6.290	3.267	3.252				
% Variance	41.9	21.8	21.7				
%Cumulative variance	41.9	63.7	85.4				

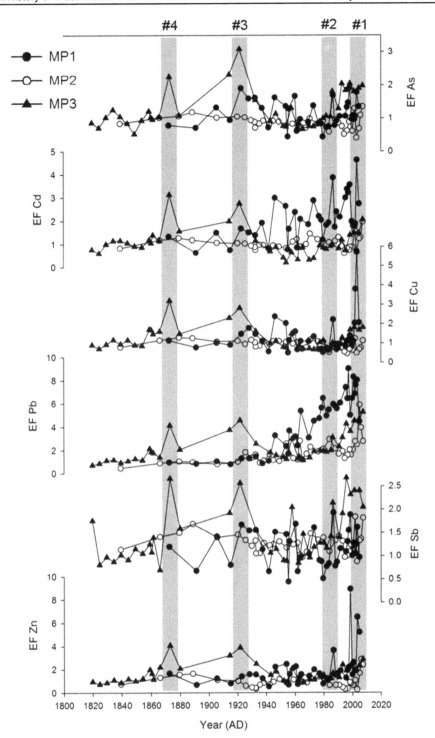

Figure 4. Temporal trends of enrichment factors (EF, see text for calculation) for As, Cd, Cu, Pb, Sb and Zn in the three cores from Motianling peatland in Great Hinggan Mountains, NE China. An EF >1 indicates that the sample is enriched, relative to background, thus four peaks of EFs are observed, representing specific historical periods (1860–1880: #4, 1920–1940: #3, and 1949-now: #1 and 2) of society development in China.

3.3. Principal Component Analysis

The statistical identification of the main factors that control the elemental distribution in peat has been based on PCA of the complete dataset, using a varimax with Kaiser normalization rotation method. Three principle components have Eigen-values >1 and explain 85% of the total variance of element concentrations in the three cores (Table 4). For all elements, at least 68% or more of their variance is explained by the exacted components (Table 4). The first component (PC1) accounts for a large proportion of the variance (41.9%) and is characterized by the high positive loadings (0.777–0.941) of Ba, Co, Cr, Mo, Ni, Sr and Tl (Table 4). The second component (PC2) accounts for 21.8% of the total variance and is dominated by the high positive loadings (0.865–0.957) of As, Cu and Sb, and a moderate positive loading of Cd (0.535). The third component (PC3) accounts for 21.7% of the total variance and is characterized by the high negative loading of the $^{206}Pb/^{207}Pb$ ratio (−0.894), the high positive loadings of the $^{208}Pb/^{206}Pb$ (0.888) and Pb (0.789), and the moderate positive loading of Zn (0.561). It is worthy to note that Cd also has a moderate positive loading (0.455) associated to PC1 and a moderate positive loading (0.516) associated to PC3, and Zn also has a moderate positive loading (0.512) associated to PC1.

3.4. Enrichment Factor Analysis

Combined with the established local pre-industrial backgrounds, enrichment factors were calculated through normalizing the measured heavy metal content (As, Cd, Cu, Pb, Sb and Zn) with respect to a reference metal (*i.e.*, Ti) for metal concentrations in the ombrotrophic upper profiles. Their temporal variations were shown in Figure 4 for the three cores. Given an EF > 1 suggesting that the sample is enriched relative to background, four peaks of EFs were supposed to be in the following order: #1 in the 2000s, #2 in the 1980s, #3 in the 1930s and #4 in the 1870s.

4. Discussion

4.1. The Ombrotrophic vs. Minerotrophic Character of the Peat

The ombrotrophic peats receive their inorganic minerals predominantly from the atmosphere and are mainly composed of *Sphagnum* and sedge remains, but minerotrophic peats, as the name suggests, are also fed by surface runoff and groundwaters and are relatively rich in inorganic components [14]. Motianling peatland is about 2.6 km from the southwestern bank of Da'erbing Lake which is a lava-dammed lake with an altitude of *ca.* 1200 m (Figure 1B). A large number of types of mires are widely distributed on lakeside, footslope as well as hillside, with a transect of eutrophic, mesotrophic and oligotrophic mires from the lakeside up to the hillside [63]. The rain-fed oligotrophic peatland on the hillside is underlain by the permafrost of the basalt weathering crust and predominantly consists of mosses (*Sphagumacutifolium, Sphagnum girgensohnii*), ericaceous shrubs (*Ledumpalustre var. angustum, Vacciniumvitis-idaea*), and a sparse pine cover (*Pinuspumila, Larixgmelini*) [26,63]. Physicochemical properties of the peat (*i.e.*, DBD, WAT, ASH, TOC and TN) as well as other proxies reported elsewhere [28] have indicated its ombrotrophic character in the upper sections (above 45 cm for MP1, 50 cm for MP2 and 60 cm for MP3) and minerotrophic character in the lower sections. More evidence from the degree of peat decomposition, pH and grain-size composition can be found here to support such structural characterization of peat. In the lower sections, peats are much more decomposed than the above sections as shown by the C/N ratios (Figure 2) and have a higher pH value (averaged 4.6, 4.8 and 4.9 for the upper sections of MP1, MP2 and MP3; 5.1, 5.3 and 5.3 for the corresponding lower sections). Furthermore, the homogeneous fine grain minerals reflected the sole possibility of the plant cover in the area. At present, the dominant plant is *Sphagnum*, so it could be inferred that the *Sphagnum* had grown in this area, and thus formed a peat bog. With the development of peatland, the deeper sections of peats with high ash content and large degree of decomposition were probably affected by thawing of the frozen soil in the bottom of cores and they have a minerotrophic nature.

Stable Pb isotope ratios were used for the first time ever to indicate the ombrotrophic *versus* minerotrophic system with the logic that a high ratio of ^{206}Pb/^{207}Pb ($\approx$1.2, close to the typical crustal soil [64]) could fingerprint that lithogenic components contribute more to the Pb content in the peat at the base of the individual profiles. Here, in the lower section of each core, ratios of ^{206}Pb/^{207}Pb fluctuate slightly around 1.19 (1.1912 ± 0.0018 (n = 9) for MP1, 1.1896 ± 0.0024 (n = 8) for MP2, and 1.1868 ± 0.0032 (n = 10)) and are very close to the signature of the typical of crustal soil, which is just corresponding to the boundary of minerotrophic and ombrotrophic peats (Figure 5). In addition, the temporal variations of ^{206}Pb/^{207}Pb ratio are consistent with the ash content profiles for three peat cores (r = 0.624, P < 0.001, n = 114). Therefore, we think these underlying peats show the most lithogenic signatures. In addition, previous studies also showed that these peats were affected by the effects of permafrost [56] or the addition of volcanic ash during peat formation process [65]. All in all, the boundary partition above is reasonable and ombrotrophic peat is appropriate to be used as a record of atmospheric metal deposition in northeast China.

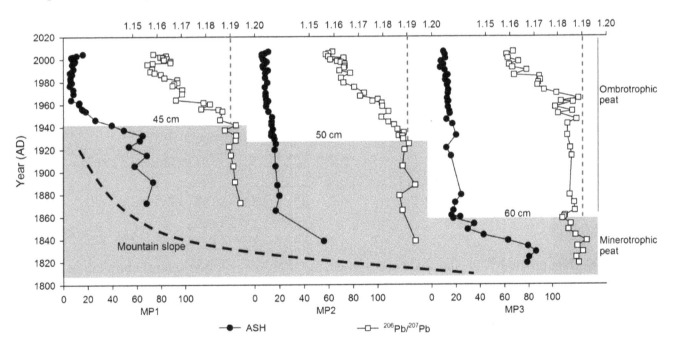

Figure 5. Temporal profiles of ^{206}Pb and ^{207}Pb ratios and their comparisons with ash content ofthree core samples from Motianling peatland in Great Hinggan Mountains, NE China. Sections below 45 cm (MP1), 50 cm (MP2), and 60 cm (MP3) are attributed to minerotrophic peat and marked by a grey dashed box.

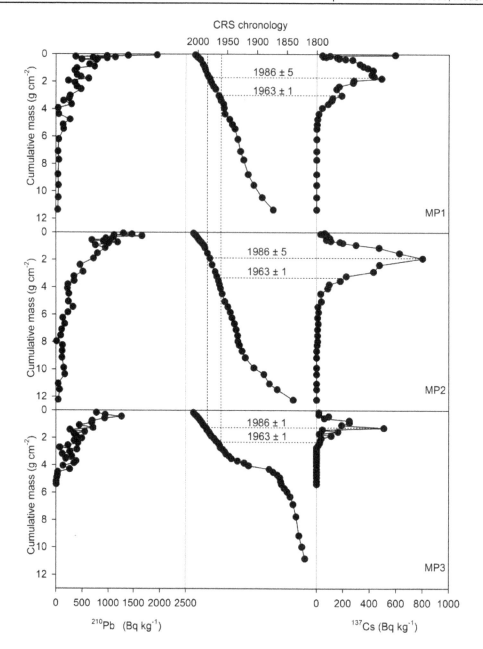

Figure 6. Unsupported ^{210}Pb and ^{137}Cs activity (Bq·kg^{-1}) profiles plotted against cumulative mass depth (g·cm^{-2}) in the three cores from Motianling peatland in Great Hinggan Mountains, northeast China, and comparison of the chronology estimated by ^{210}Pb CRS model with the ^{137}Cs time marker.

4.2. Validation of the Peat Age

Depth profiles for ^{210}Pb and ^{137}Cs activities have been plotted with peat depth in a normal scale [28], which did not take into account the influence of compaction. The major ^{137}Cs peak was assigned to the weapon's testing peak *ca.* 1963 in the two previous reports [26,28]. This leads to the conclusion that the ^{137}Cs peak dates have a *ca.* 20-yr younger age than ^{210}Pb chronology, although the average PAR derived from both ages for the past 45 years is in good agreement. As a result, the ^{210}Pb and ^{137}Cs activities were re-plotted against cumulative peat mass (g·cm^{-2}) (Figure 6). Unsupported ^{210}Pb activities generally show decreasing trend with depth from the surface of the peat cores until negligible concentrations of the unsupported ^{210}Pb were attained at depth of 6.22 g·cm^{-2} (corresponding to 51 cm) for MP1, 11.06 g·cm^{-2} (59 cm) for MP2 and 5.19 g·cm^{-2} (59 cm) for MP3. Three ^{210}Pb chronologies were calculated using the constant rate of supply model (CRS) and they are validated by ^{137}Cs distribution (Figure 6). The ^{137}Cs profiles reflect their historical fallout and exhibit an obvious subsurface peak with maximum activity at depth of 1.78 g·cm^{-2} (23 cm) for MP1, 1.91 g·cm^{-2} (17 cm) for MP2, and 1.28 g·cm^{-2} (15 cm) for MP3. A smaller peak is detected in deeper layers, 3.03 g·cm^{-2} (33 cm) for MP1, 3.81 g·cm^{-2} (27 cm) for MP2, and 2.56 g·cm^{-2} (31 cm) for MP3. The major subsurface peaks here were corresponding to 1986 and the smaller peaks below them were corresponding to 1963, because widespread fallout of ^{137}Cs in the Northern Hemisphere after the Chernobyl reactor disaster in 1986 is more likely to form the subsurface peak with maximum activity in ^{137}Cs profile [66]. Significant fallout of ^{137}Cs after the Chernobyl accident was also reported in the Xiaolongwan lake sediment in Jilin province, northeast China [67] and in the coastal wetland sediment of the northern Beibu Gulf, South China Sea [68]. Such time markers are very consistent with the ^{210}Pb chronology within a five-year deviation (Figure 6). In addition, the ^{210}Pb ages were also validated by ^{14}C dates from another core. Lin *et al.* [63] reported three ^{14}C ages for a 106 cm long core from the same peatland and the calibrated ^{14}C age (2σ) at 75 cm depth layer is 176 y·cal.before present (BP) (170 ± 60 y·BP). This is in good agreement with our ^{210}Pb age, 190 ± 20 y·cal. BP at 78 cm depth, which elucidates that our chronologies established from ^{210}Pb and ^{137}Cs techniques are coherent and reliable.

4.3. Geochemical Variability of the Peat Records Derived from PCA Analysis

The first component (PC1) clustered typical lithogenic elements of Ba and Sr [69,70], and also TMs of Co, Cr, Mo, Ni and Tl; thus, PC1 seems to reflect the geogenic contribution. The second component (PC2) clustered typical pollution elements of As, Cd, Cu and Sb, and probably represented a long distance source because the enrichment of As was interpreted as long-range signal by Muller *et al.* [51]. The third component (PC3) clustered Zn, Pb and its isotopes, and PC3 seems to be associated with local anthropogenic source from leaded gasoline burning and from ore mining and processing such as the Bairendaba Ag-Pb-Zn deposit (as shown in Figure 1), one of the largest polymetallic deposits in this region [71]. It is worthy noting that Cd also has a moderate loading (0.455) associated to PC1 and a moderate loading (0.516) associated to PC3, and Zn also has a moderate positive loading (0.512) associated to PC1.Therefore, both Cd and Zn would also be affected by the geogenic contribution (20.7% and 26.2%), and 26.6% of Cd would be derived from industrial manufacturing and fossil-fuel combustion.

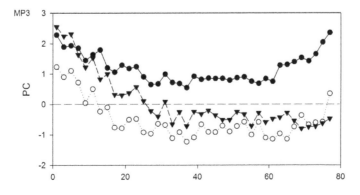

Figure 7. *Cont.*

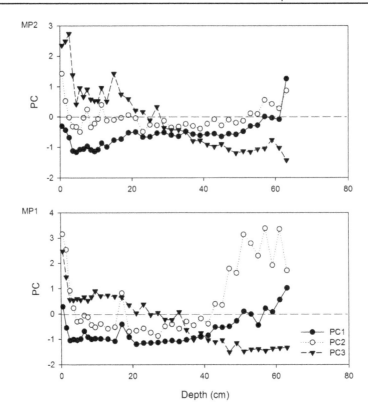

Figure 7. Variation of principal component scores (PC1, PC2 and PC3) against depth for the three peat cores from Great Hinggan Mountain, northeast China.

The records of factor scores for each component are presented in Figure 7. As stated in Cortizas *et al.* [49], factor scores are non-dimensional, average centered values and can measure the intensity of each principal component. In the surface or near-surface (*ca.* 5 cm depth of MP1 and MP2, and 25 cm depth of MP3) of the cores, all three components showed an increasing trend with depth upward, indicating increasing trends of mineral dust content and atmospheric metal pollution in the peat. In the lower sections, PC3 showed relative stable and significant correlations ($P < 0.001$) between cores, $r = 0.902$ (MP1 and MP2), 0.747 (MP1 and MP3) and 0.848 (MP2 and MP3), respectively. This was mainly a result of the continuous and stable local anthropogenic pollutants input. However, the other two components showed quite similar and large variations with depth, which was probably due to the long-distance contributions of pollution aerosol and dust. In a previous study [28], the typical lithogenic element (Ti) was used to successfully reconstruct recent deposition of atmospheric soil dust (ASD). The significant correlations between factor scores of PC1 and Ti concentration, ash content and ASD suggest that PC1 is indicative of natural/geogenic sources (Figure 8). Variations of PC1 are similar as suggested by their significant correlations between MP1 and MP2 ($r = 0.672$, $P < 0.001$) and between MP1 and MP3 ($r = 0.397$, $P = 0.015$). The large values of PC1 and PC2 at the bottom of cores would be resulted from the sediment/soil character of the peat, which is also in agreement with the fact that part of the variation in the concentrations of Cd is associated with geogenic sources.

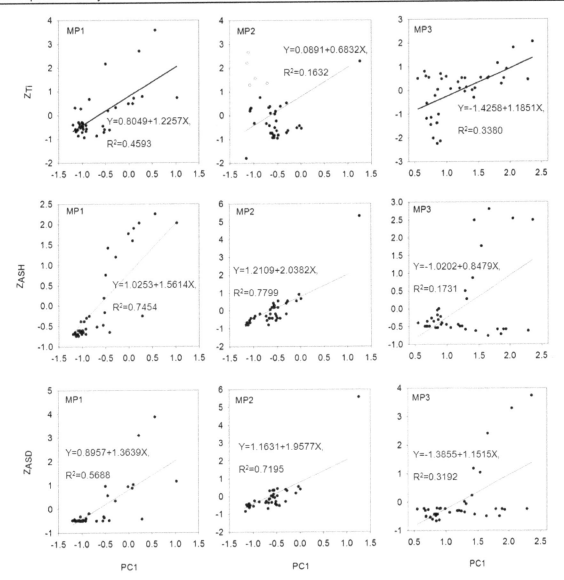

Figure 8. Correlations between factor scores of PC1 (indicative of natural/geogenic sources) and Ti concentration, Ash content (ASH) and the calculated depositional rate of atmospheric soil dust (ASD). Data of Ti, ASH and ASD are from Bao *et al.* [28] and are in normalized formation. For regression of PC1 against Ti in MP2 core, five abnormal high values of Ti in the near-surface (at 3.5 cm, 7.5–9.5 cm and 15 cm depth) are excluded and indicated by a white dot.

4.4. Historical Trends of Accumulation Rates for the Pollution Metals

Assuming the relationship between peat accumulation and the age explained before [28], we can assign each peat section an estimated age and determine the ARs' evolution of those elements (Figure 9). There were large AR values for the metals in the minerotrophic sections of the bottom cores, and this pattern is consistent with the vertical distribution of elements (Figure 3), which would be controlled by the large abundance of mineral matter. The ARs of these elements in these sections could not represent the external atmospheric fluxes of metals deposition and were excluded using different color of dots for those points.

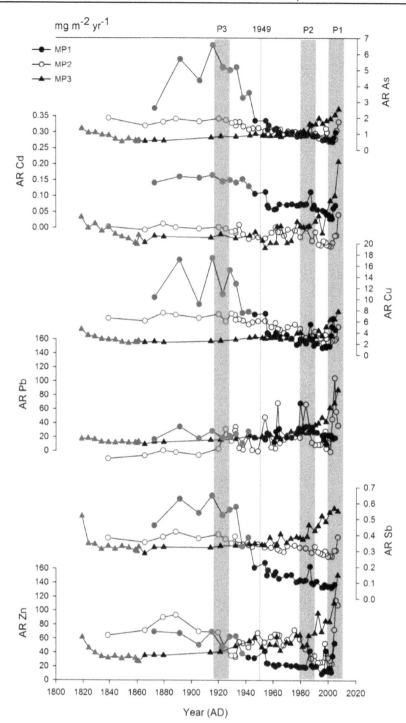

Figure 9. Temporal variation of accumulation rate (AR, mg·m^{-2}·yr^{-1}) of typical pollution elements (As, Cd, Cu, Pb, Sb and Zn) identified by PCA for three cores from Motianling peatland in Great Hinggan Mountains, northeast China. The minerotrophic sections (*i.e.*, fen) below 45 cm for MP1, 50 cm for MP2 and 60 cm for MP3 are marked by red dots. The year of New China foundation in 1949 was indicated by a grey line. Three peaks of ARs for these metals were observed since the 2000s (P1), in the 1980s (P2) and in the 1930s (P3).

For the above ombrotrophic sections, an increasing period of ARs for these metals since the 2000s (p1) and a peak of ARs in the 1980s (p2) were observed in the three cores. Both periods are corresponding to the variation of EFs in #1 and #2. Starting in 1949 through the 1980s, there was huge industrial and agricultural efforts for economic development due to government policies, for example, the Great Leap Forward in 1958–1960. Since 1978, China pursued relatively open economic policies (*i.e.*, Reform and Opening up) and has entered the most robust stage of its industrial revolution and economic growth. Total energy consumption and coal consumption increased by 15.1% from 1952–1990 and by 22.9% from 1990–2010; emissions of industrial pollutants increased annually by 14.9% from 1952–1990 and by 12.2% from 1990–2010; the proportion of total pollution emitted by industries in China was 42.6% from 1952–1990 and 43.6% from 1990–2010 [72]. Therefore, the consistency of ARs and EFs support the fact that recent industrial development since Reform and Opening up in 1978 (p2) and modern large-scale urbanization associated with increasing vehicle emissions (p1) have led to environmental pollution and thus elevated records of these trace metals in peat.

In addition, it is worthy noting the period in the 1930s, although it was located in the minerotrophic section of MP1. The elevated ARs TMs, especially for Pb, were observed in the period (p3) for MP2 and MP3 cores. This characterization only corresponds to the variation of EFs in the same period (#3), which occurred during the long-term war periods before the foundation of New China and would have resulted from pollutants due to combustions of fuel and ecological damage due to deforestation and energy exploitation in China. The peak in the 1870s (#4) would be explained by the onset of industrial revolution in China and extensive human migration and cultivation in northeast China in the late Qing dynasty [73]; however, this character was not found from ARs whose enrichment character would be probably offset by variation of PAR due to great amount of minerals. Therefore, these periods correspond quite well to the specific social stages during the development history of China.

4.5. Sourcing the TMs Pollution by Pb Isotopic Composition

The temporal variations of ^{206}Pb/^{207}Pb ratio are shown in Figure 10a, and exhibit a clear trend to atmospheric Pb pollution in the past 150 years, with the highest concentration of Pb found in recent sediments and associated with less radiogenic (more contaminated) values of ^{206}Pb/^{207}Pb ratios. The phase-out of leaded gasoline in China started in the 1990s, with leaded gasoline being banned in several major cities (*i.e.*, Beijing and Shanghai) in 1997 and 1998, before a nationwide ban in 2000 [74]. Our results show that there is still Pb contamination in our rural mountainous site despite the leaded gasoline phase-out. This is consistent with the conclusion from monitoring data which indicated that the Pb concentrations remained relatively high in aerosol samples in Shanghai and Tianjing after the phase-out of leaded gasoline [75,76]. In order to assess Pb contamination and identify potential Pb sources of peats, the correlation between the ^{206}Pb/^{207}Pb ratio *vs.* 1/Pb concentrations were analyzed (Figure 10b). The significant linear trends represented a mixture between two main sources of natural geogenic Pb and anthropogenic pollution Pb [27]. Three-isotope (^{206}Pb/^{207}Pb and ^{208}Pb/^{206}Pb) diagrams were built with an attempt to assign an origin to the anthropogenic sources (Figure 10c). The comparisons between Pb isotopic ratios in peats and other environmental samples show that the ^{206}Pb/^{207}Pb ratios in the minerotrophic sections fall in a tight cluster representing the natural source; those ratios in the ombrotrophic sections display a relatively linear feature, covering those in northern China coal [77] and aerosol samples in Beijing, China and Ulanbaator, Mongolia [78]. This spread most probably reflected a mixture of emissions from the burning of petrol and coal. Mongolia is not far and our study site is located in the down wind direction as shown by the wind frequency (Figure 1). It has been reported that soil mineral dust input for the Motianling peat bog came mainly from the long-range sources in northern China and Mongolia [28]. Thus, aerosols from Mongolia were also regarded as mainly contributing to deposited Pb. In northeast China, coal is one of the most important energy resources in urban economical and industrial development, and the leaded particular matter from coal combustion could contribute to other parts of anthropogenic Pb. Unfortunately, there is not very much information on coal, local ores, and alkyl lead ores used in both countries as well as local aerosol data. It is therefore not possible to apportion a reasonable contribution from each source to anthropogenic Pb in the peat bog currently.

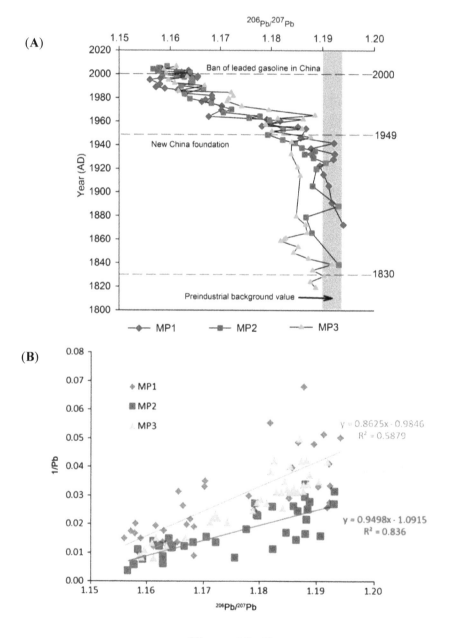

Figure 10. *Cont.*

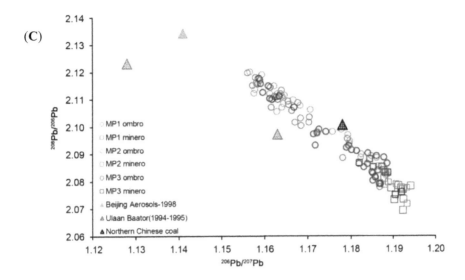

Figure 10. (A) Chronological variation of $^{206}Pb/^{207}Pb$ *vs.* peat growth year derived from ^{210}Pb technique. The period prior to 1830 is considered as pre-industrial and the background values larger than 1.19 in this study are included in the grey box. Two periods of anthropogenic Pb contributions are identified by the time markers of the foundation of New China in 1949 and the leaded gasoline phase-out in China nationwide in 2000; (B) Diagram of $^{206}Pb/^{207}Pb$ *vs.* 1/Pb. They exhibit significant correlations and the large amounts of Pb correspond to the lower ratio of ^{206}Pb and ^{207}Pb; (C) Three isotope plots ($^{206}Pb/^{207}Pb$ *vs.* $^{208}Pb/^{206}Pb$) diagram with the least radiogenic samples (circle) corresponding to the ombrotrophic peat and the more radiogenic samples (square) corresponding to the minerotrophic peats in Motianling. Values for aerosols samples collected in 1998 at Beijing and from 1994–1995 at Ulaan Baator are from Bollhofer and Rosman [78] and for northern Chinese coal are from Mukai *et al.* [77] (triangle).

5. Conclusions

This study reported a high resolution record of TMs over the last 200 years derived from multiple ^{210}Pb-dated cores collected from an ombrotrophic peatland in Great Hinggan Mountains, northeast China. The general agreement in the elemental concentration profiles suggested that all investigated elements were conserved in the Motianling bog. This study suggested a necessity to take more than one core to show the heterogeneity in peat accumulation and to investigate the utility of peat as an archive of TMs. AR TMs, EF, PCA and Pb isotope analyses were used to identify the pre-industrial background of typical pollution elements and the impact and trends in anthropogenic contributions since the onset of the Industrial Revolution at 1830 AD. For the typical pollution elements identified by PCA (As, Cd, Cu, Pb, Sb and Zn), the maximum values of AR were found near the surface of the peatland (in peat layers dated from the 1980s), and an increasing trend of AR for these metals was observed since the 2000s. The Pb isotopic composition shows a clear anthropogenic signature probably from re-emission of Pb gasoline contaminated soil particles or from a source with a similar isotopic source, even after leaded gasoline phase-out in 2000. The major potential sources of anthropogenic Pb pollution are likely the emissions from coal combustion in northern China and the aerosols from leaded gasoline burning in the local area as well as long-distance transportation from Mongolia. The historical profile of TMs in the

Motianling peatland is in agreement with industrial development in China, reflecting the influence of regional environmental pollution.

Acknowledgments

We thank Yuxia Zhang and Haiyang Zhao for sampling assistance and Yuxin Zhu and Weilan Xia for valuable help during laboratory analysis. We are grateful to three anonymous referees for their useful comments on the paper. This research was funded by the National Natural Science Foundation of China (no.41301215), the Natural Science Foundation of Jiangsu Province, China (no.BK20131058), and the National Basic Research Program of China (no.2012CB956100).

Author Contributions

Kunshan Bao and Guoping Wang conceived and designed the experiments, Kunshan Bao and Ji Shen collected the data, Kunshan Bao and Gaël Le Roux analyzed the data, Kunshan Bao, Ji Shen and Guoping Wang wrote the paper, and all authors contributed to discussions of the manuscript and approved the final manuscript.

Conflicts of Interest

The authors declare no conflict of interest.

References

1. Nriagu, J.O. A history of global metal pollution. *Science* **1996**, *272*, 223–224.
2. Pacyna, J.M.; Pacyna, E.G. An assessment of global and regional emissions of trace metals to the atmosphere from anthrogpogenic sources worldwide. *Environ. Rev.* **2001**, *9*, 269–298.
3. Ilyin, I.; Rozovskaya, O.; Travnikov, O.; Varygina, M.; Aas, W. Heavy Metals: Transboundary Pollution of the Environment. Available online: http://emep.int/publ/common_publications.html (accessed on 12 March 2015).
4. Le Roux, G.; Aubert, D.; Stille, P.; Krachler, M.; Kober, B.; Cheburkin, A.; Bonani, G.; Shotyk, W. Recent atmospheric Pb deposition at a rural site in southern Germany assessed using a peat core and snowpack, and comparison with other archives. *Atmos. Environ.* **2005**, *39*, 6790–6801.
5. Bindler, R.; Yu, R.; Hansson, S.; Claben, N.; Karlsson, J. Mining, metallurgy and the historical origin of mercury pollution in lakes and watercourses in Central Sweden. *Environ. Sci. Technol.* **2012**, *46*, 7984–7991.
6. Martinez Cortizas, A.; Peiteado Varela, E.; Bindler, R.; Biester, H.; Cheburkin, A. Reconstructing historical Pb and Hg pollution in NW Spain using multiple cores from the Chao de Lamoso bog (Xistral Mountains). *Geochim. Cosmochim. Acta* **2012**, *82*, 68–78.
7. Gallego, J.L.R.; Ortiz, J.E.; Sierra, C.; Torres, T.; Llamas, J.F. Multivariate study of trace element distribution in the geological record of Ronanzas peat bog (Asturias, N. Spain). Paleoenvironmental evolution and human activities over the last 8000 cal y BP. *Sci. Total Environ.* **2013**, *454*, 16–29.

8. Kuttner, A.; Mighall, T.M.; de Vleeschouwer, F.; Mauquoy, D.; Martinez Cortizas, A.; Foster, I.D.L.; Krupp, E. A 3300-year atmospheric metal contamination record from Raeburn Flow raised bog, south west Scotland. *J. Archaeol. Sci.* **2014**, *44*, 1–11.

9. Hong, S.M.; Candelone, J.P.; Patterson, C.C.; Boutron, C.F. Greenland ice evidence of hemispheric lead pollution two millennia ago by Greek and Roman civilizations. *Science* **1994**, *265*, 1841–1843.

10. Renberg, I.; Wik-Persson, M.; Emteryd, O. Pre-industrial atmospheric lead contamination detected in Swedish lake sediments. *Nature* **1994**, *368*, 323–326.

11. Rydberg, J.; Martinez Cortizas, A. Geochemical assessment of an annually laminated lake sediment record from northern Sweden: A multi-core, multi-element approach. *J. Paleolimnol.* **2014**, *51*, 499–514.

12. Shotyk, W.; Weiss, D.; Appleby, P.; Cheburkin, A.; Frei, R.; Gloor, M.; Kramers, J.D.; Reese, S.; van der Knaap, W. History of atmospheric lead deposition since 12,370 ^{14}C y BP from a peat bog, Jura Mountains, Switzerland. *Science* **1998**, *281*, 1635–1640.

13. Allan, M.; le Roux, G.; de Vleeschouwer, F.; Bindler, R.; Blaauw, M.; Piotrowska, N.; Sikorski, J.; Fagel, N. High-resolution reconstruction of atmospheric deposition of trace metals and metalloids since AD 1400 recorded by ombrotrophic peat cores in Hautes-Fagnes, Belgium. *Environ. Poll.* **2013**, *178*, 381–394.

14. Shotyk, W. Peat bog archives of atmospheric metal deposition: Geochemical evaluation of peat profiles, natural variations in metal concentrations, and metal enrichment factors. *Environ. Rev.* **1996**, *4*, 149–183.

15. Shotyk, W.; Weiss, D.; Kramers, J.D.; Frei, R.; Cheburkin, A.K.; Gloor, M.; Reese, S. Geochemistry of the peat bog at Etang de la Gruère, Jura Mountains, Switzerland, and its record of atmospheric Pb and lithogenic trace metals (Sc, Ti, Y, Zr, and REE) since 12,370 ^{14}C y BP. *Geochim. Cosmochim. Acta* **2001**, *65*, 2337–2360.

16. Bindler, R. Mired in the past-looking to the future: Geochemistry of peat and the analysis of past environmental changes. *Glob. Planet. Change* **2006**, *53*, 209–221.

17. Damman, A.W.H. Hydrology, development, and biogeochemistry of ombrogenous peat bog with special reference to nutrient relocation in a western Newfoundland bog. *Can. J. Bot. Sci.* **1986**, *64*, 384–394.

18. Shotyk, W.; Krachler, M.; Martinez Cortizas, A.; Cheburkin, A.; Emons, H. A peat bog record of natural, pre-anthropogenic enrichments of trace elements in atmospheric aerosols since 12,370 ^{14}C y BP, and their variation with Holocene climate change. *Earth Planet. Sci. Lett.* **2002**, *199*, 21–37.

19. Madsen, P. Peat bog records of atmospheric mercury deposition. *Nature* **1981**, *293*, 127–130.

20. Martinez Cortizas, A.; Pontevedra-Pombal, X.; Garcia-Rodeja, E.; Novoa-Munoz, J.C.; Shotyk, W. Mercury in a Spanish peat bog: Archive of climate change and atmospheric metal deposition. *Science* **1999**, *284*, 939–942.

21. Shotyk, W.; Goodsite, M.E.; Roos-Barraclough, F.; Frei, R.; Heinemeier, J.; Asmund, G.; Lohse, C.; Hansen, T.S. Anthropogenic contributions to atmospheric Hg, Pb and As accumulation recorded by peat cores from southern Greenland and Denmark dated using the ^{14}C "bomb pulse curve". *Geochim. Cosmochim. Acta* **2003**, *67*, 3991–4011.

22. Coggins, A.M.; Jenning, S.G.; Ebinghaus, R. Accumulation rates of the heavy metals lead, mercury and cadmium in ombrotrophic peatlands in the west of Ireland. *Atmos. Environ.* **2006**, *40*, 260–278.

23. Pratte, S.; Mucci, A.; Garneau, M. Historical records of atmospheric metal deposition along the St. Lawrence Valley (eastern Canada) based on peat bog cores. *Atmos. Environ.* **2013**, *79*, 831–840.

24. Shi, W.F.; Feng, X.B.; Zhang, G.; Ming, L.L.; Yin, R.S.; Zhao, Z.Q.; Wang, J. High-precision measurement of mercury isotope ratios of atmospheric deposition over the past 150 years recorded in a peat core taken from Hongyuan, Sichuan Province, China. *Chin. Sci. Bull.* **2011**, *56*, 877–882.

25. Tang, S.; Huang, Z.; Liu, J.; Yang, Z.; Lin, Q. Atmospheric mercury deposition recorded in an ombrotrophic peat core from Xiaoxing'an Mountain, Northeast China. *Environ. Res.* **2012**, *118*, 145–148.

26. Bao, K.; Xia, W.; Lu, X.; Wang, G. Recent atmospheric lead deposition recorded in an ombrotrophic peat bog of Great Hinggan Mountains, Northeast China, from ^{210}Pb and ^{137}Cs dating. *J. Environ. Radioact.* **2010**, *101*, 773–779.

27. Ferrat, M.; Weiss, D.; Dong, S.; Large, D.J.; Spiro, B.; Sun, Y.; Gallagher, K. Lead atmospheric deposition rates and isotopic trends in Asian dust during the last 9.5 kyr recorded in an ombrotrophic peat bog on the eastern Qinghai-Tibetan Plateau. *Geochim. Cosmochim. Acta* **2012**, *82*, 4–22.

28. Bao, K.; Xing, W.; Yu, X.; Zhao, H.; McLaughlin, N.; Lu, X.; Wang, G. Recent atmospheric dust deposition in an ombrotrophic peat bog in Great Hinggan Mountain, Northeast China. *Sci. Total Environ.* **2012**, *431*, 33–45.

29. Ferrat, M.; Weiss, D.; Spiro, B.; Large, D.J. The inorganic geochemistry of a peat deposit on the eatern Qinghai-Tibetan Plateau and insights into changing atmospheric circulation in central Asia during the Holocene. *Geochim. Cosmochim. Acta* **2012**, *91*, 7–31.

30. Grybos, M.; Davranche, M.; Gruau, G.; Petitjean, P. Is trace metal release in wetland soils controlled by organic matter mobility or Fe-oxyhydroxides reduction? *Int. J. Coal Geol.* **2007**, *314*, 490–501.

31. Rausch, N.; Ukonmaanaho, L.; Nieminen, T.; Krachler, M.; Shotyk, W. Porewater evidence of metal (Cu, Ni, Co, Zn, Cd) mobilization in an acidic, ombrotrophic bog impacted by a smelter, Harjavalta, Finland and comparison with references sites. *Environ. Sci. Technol.* **2005**, *39*, 8207–8213.

32. Biester, H.; Hermanns, Y.M.; Martinez Cortizas, A. The influence of organic matter decay on the distribution of major and trace elements in ombrotrophic mires—A case study from the Harz Mountains. *Geochim. Cosmochim. Acta* **2012**, *84*, 126–136.

33. MacKenzie, A.B.; Logan, E.M.; Cook, G.T.; Pulford, I.D. Distribution, inventories and isotopic composition of lead in ^{210}Pb-dated peat cores from contrasting biogeochemical environments: Implication for lead mobility. *Sci. Total Environ.* **1998**, *223*, 25–35.

34. Le Roux, G.; Weiss, D.; Grattan, J.; Givelet, N.; Krachler, M.; Cheburkin, A.; Rausch, N.; Kober, B.; Shotyk, W. Identifying the sources and timing of ancient and medieval atmospheric lead pollution in England using a peat profile from Lindow bog, Manchester. *J. Environ. Monit.* **2004**, *6*, 502–510.

35. Mihaljevic, M.; Zuna, M.; Ettler, V.; Sebek, O.; Strnad, L.; Golias, V. Lead fluxes, isotopic and concentration profiles in a peat deposit near a lead smelter (Pribram, Czech Republic). *Sci. Total Environ.* **2006**, *372*, 334–344.

36. Weiss, D.; Shotyk, W.; Kramers, J.D.; Gloor, M. Sphagnum mosses as archives of recent and past atmospheric lead deposition in Switzerland. *Atmos. Environ.* **1999**, *33*, 3751–3763.

37. Shotyk, W. Review of the inorganic geochemistry of peats and peatland waters. *Earth Sci. Rev.* **1988**, *25*, 95–176.

38. Novak, M.; Pacherova, P. Mobility of trace metals in pore waters of two Central European peat bogs. *Sci. Total Environ.* **2008**, *394*, 331–337.

39. Nieminen, T.; Ukonmaanaho, L.; Shotyk, W. Enrichments of Cu, Ni, Zn, Pb and As in an ombrotrophic peat bog near a Cu-Ni smelter in Southwest Finland. *Sci. Total Environ.* **2002**, *292*, 81–89.

40. Shotbolt, L.; Buker, P.; Ashmore, M.R. Reconstructing temporal trends in heavy metal deposition: Assessing the value of herbarium moss samples. *Environ. Poll.* **2007**, *147*, 120–130.

41. Rausch, N.; Nieminen, T.; Ukonmaanaho, L.; le Roux, G.; Krachler, M.; Cheburkin, A.; Bonani, G.; Shotyk, W. Comparison of atmospheric deposition of copper, nickel, cobalt, zinc, and cadmium recorded by Finnish peat cores with monitoring data and emission records. *Environ. Sci. Technol.* **2005**, *39*, 5989–5999.

42. Krachler, M.; Mohl, C.; Emons, H.; Shotyk, W. Atmospheric deposition of V, Cr, and Ni since the late glacial: Effects of climatic cycles, human impacts, and comparison with crustal abundances. *Environ. Sci. Technol.* **2003**, *37*, 2658–2667.

43. Ukonmaanaho, L.; Nieminen, T.; Rausch, N.; Shotyk, W. Heavy metal and arsenic profiles in ombrotrophic peat cores from four differently loaded areas in Finland. *Water Air Soil Poll.* **2004**, *58*, 277–294.

44. Shotyk, W.; Krachler, M.; Chen, B. Antimony in recent, ombrotrophic peat from Switzerland and Scotland: Comparison with natural background values (5320 to 8020 [14]C y BP) and implications for the global atmospheric Sb cycle. *Glob. Biogeochem. Cycles* **2004**, *18*, GB1016.

45. Cloy, J.M.; Farmer, J.G.; Graham, M.C.; MacKenzie, A.B. Retention of As and Sb in ombrotrophic peat bogs: Records of As, Sb, and Pb deposition at four Scottish sites. *Environ. Sci. Technol.* **2009**, *43*, 1756–1762.

46. Rothwell, J.J.; Taylor, K.G.; Chenery, S.R.N.; Cundy, A.B.; Evans, M.G.; Allott, T.E.H. Storage and behavior of As, Sb, Pb, and Cu in bmbrotrophic peat bogs under contrasting water table conditions. *Environ. Sci. Technol.* **2010**, *44*, 8497–8502.

47. Kylander, M.E.; Bindler, R.; Cortizas, A.M.; Gallagher, K.; Mörth, C.-M.; Rauch, S. A novel geochemical approach to paleorecords of dust deposition and effective humidity: 8500 years of peat accumulation at Store Mosse (the "Great Bog"), Sweden. *Quatern. Sci. Rev.* **2013**, *69*, 69–82.

48. Margalef, O.; Cañellas-Boltà, N.; Pla-Rabes, S.; Giralt, S.; Pueyo, J.J.; Joosten, H.; Rull, V.; Buchaca, T.; Hernández, A.; Valero-Garcés, B.L.; *et al.* A 70,000 year multiproxy record of climatic and environmental change from Rano Aroi peatland (Easter Island). *Glob. Planet. Change* **2013**, *108*, 72–84.

49. Martinez Cortizas, A.; Lopez-Merino, L.; Bindler, R.; Mighall, T.; Kylander, M. Atmospheric Pb pollution in N Iberia during the late Iron Age/Roman times reconstructed using the high-resolution record of La Molina mire (Asturias, Spain). *J. Paleolimnol.* **2013**, *50*, 71–86.

50. Gao, C.; Bao, K.; Lin, Q.; Zhao, H.; Zhang, Z.; Xing, W.; Lu, X.; Wang, G. Characterizing trace and major elemental distribution in late Holocene in Sanjiang Plain, Northeast China: Paleoenvironmental implications. *Quatern. Int.* **2014**, *349*, 376–383.

51. Muller, J.; Kylander, M.; Martinez-Cortizas, A.; Wüst, R.A.J.; Weiss, D.; Blake, K.; Coles, B.; Garcia-Sanchez, R. The use of principle component analyses in characterising trace and major elemental distribution in a 55 kyr peat deposit in tropical Australia: Implications to paleoclimate. *Geochim. Cosmochim. Acta* **2008**, *72*, 449–463.

52. Shotyk, W.; Blaser, P.; Grunig, A.; Cheburkin, A.K. A new approach for quantifying cumulative, anthropogenic, atmospheric lead deposition using peal cores from bogs: Pb in eight Swiss peat bog profiles. *Sci. Total Environ.* **2000**, *249*, 281–295.

53. Martinez Cortizas, A.; Garcia-Rodeja, E.; Pontevedra Pombal, X.; Novoa Munoz, J.C.; Weiss, D.; Cheburkin, A. Atmospheric Pb deposition in Spain during the last 4600 years recorded by two ombrotrophic peat bogs and implications for the use of peat as archive. *Sci. Total Environ.* **2002**, *292*, 33–44.

54. Kylander, M.E.; Weiss, D.J.; Cortizas, A.M.; Spiro, B.; Garcia-Sanchez, R.; Coles, B.J. Refining the pre-industrial atmospheric Pb isotope evolution curve in Europe using an 8000 year old peat core from NW Spain. *Earth Planet. Sci. Lett.* **2005**, *240*, 467–485.

55. Reimann, C.; de Caritat, P. Distinguishing between natural and anthropogenic sources for elements in the environment: Regional geochemical surveys *versus* enrichment factors. *Sci. Total Environ.* **2005**, *337*, 91–107.

56. Zhang, Y.; Niu, H. The mire in northeast of China. *Chin. Geogr. Sci.* **1993**, *3*, 238–246.

57. Bao, K.; Jia, L.; Lu, X.; Wang, G. Grain-size characteristics of sediment in Daniugou peatland in Changbai Mountains, Northeast China: Implications for atmospheric dust deposition. *Chin. Geogr. Sci.* **2010**, *20*, 498–505.

58. Taylor, S.R. Abundance of chemical elements in the continental crust: A new table. *Geochim. Cosmochim. Acta* **1964**, *28*, 1273–1285.

59. Vinogradov, A.P. Average contents of chemical elements in the major types of terrestrial igneous rocks. *Geokhimiya* **1962**, *7*, 555–571.

60. Wedephol, K.H. The composition of the continental crust. *Geochim. Cosmochim. Acta* **1995**, *59*, 1217–1232.

61. Givelet, N.; Roos-Barraclough, F.; Shotyk, W. Predominant anthropogenic sources and rates of atmospheric mercury accumulation in southern Ontario recorded by peat cores from three bogs: Comparison with natural "background" values (past 8000 years). *J. Environ. Monit.* **2003**, *5*, 935–949.

62. SPSS. *Statistical Product and Service Solution*, Version 11.5; SPSS Inc.: Chicago, IL, USA, 2002.

63. Lin, Q.; Leng, X.; Hong, B. The peat record of 1 ka of climate change in Daxing Anling. *Bull. Mineral. Petrol. Geochem.* **2004**, *23*, 15–18. (In Chinese)

64. Komarek, M.; Ettler, V.; Chrastny, V.; Mihaljevic, M. Lead isotopes in environmental sciences: A review. *Environ. Int.* **2008**, *34*, 562–577.

65. Bai, Z.; Tian, M.; Wu, F.; Xu, D.; Li, T. Yanshan, Gaoshan—Two active volcanoes of the volcanic cluster in Aershan, Inner Mongolia. *Earthq. Res. China* **2005**, *19*, 402–408.

66. Azoury, S.; Tronczyński, J.; Chiffoleau, J.F.; Cossa, D.; Nakhle, K.; Schmidt, S.; Khalaf, G. Historical records of mercury, lead, and polycyclic aromatic hydrocarbons depositions in a dated sediment core from the Eastern Mediterranean. *Environ. Sci. Technol.* **2013**, *47*, 7101–7109.

67. Xia, W.; Xue, B. Sediment rate determination by ^{210}Pb and ^{137}Cs techniques in Xiaolongwan lake in Jilin province. *Quatern. Sci.* **2004**, *24*, 124–125. (In Chinese)

68. Gan, H.; Lin, J.; Liang, K.; Xia, Z. Selected trace metals (As, Cd and Hg) distribution and contamination in the coastal wetland sediment of the northern Beibu Gulf, South China Sea. *Mar. Poll. Bull.* **2013**, *66*, 252–258.

69. Krachler, M.; Shotyk, W. Natural and anthropogenic enrichments of molybdenum, thorium, and uranium in a complete peat bog profile, Jura Mountains, Switzerland. *J. Environ. Monit.* **2004**, *6*, 418–426.

70. Shotyk, W.; Goodsite, M.E.; Roos-Barraclough, F.; Givelet, N.; le Roux, G.; Weiss, D.; Cheburkin, A.K.; Knudsen, K.; Heinemeier, J.; van der Knaap, W.O.; *et al.* Accumulation rates and predominant atmospheric sources of natural and anthropogenic Hg and Pb on the Faroe Islands. *Geochim. Cosmochim. Acta* **2005**, *69*, 1–17.

71. Ouyang, H.G.; Mao, J.W.; Santosh, M. Anatomy of a large Ag-Pb-Zn deposit in the Great Xing'an Range, Northeast China: Metallogeny associated with Early Cretaceous magmatism. *Int. Geol. Rev.* **2013**, *55*, 411–429.

72. Luo, C.; Chen, L.; Zhao, H.; Guo, S.; Wang, G. Challenges facing socioeconomic development as a result of China's environmental problems, and future prospects. *Ecol. Eng.* **2013**, *60*, 199–203.

73. Ye, Y.; Fang, X.; Khan, M.A. Migration and reclamation in northeast China in response to climatic disasters in North China over the past 300 years. *Reg. Environ. Change* **2012**, *12*, 193–206.

74. Cheng, H.; Hu, Y. Lead (Pb) isotopic fingerprinting and its applications in lead pollution studies in China: A review. *Environ. Poll.* **2010**, *158*, 1134–1146.

75. Chen, J.; Tan, M.; Li, Y.; Zhang, Y.; Lu, W.; Tong, Y.; Zhang, G.; Li, Y. A lead isotope record of Shanghai atmospheric lead emissions in total suspended particles during the period of phasing out of leaded gasoline. *Atmos. Environ.* **2005**, *39*, 1245–1253.

76. Wang, W.; Liu, X.; Zhao, L.; Guo, D.; Tian, X.; Adams, F. Effectiveness of leaded petrol phase-out in Tianjin, China based on the aerosol lead concentration and isotope abundance ratio. *Sci. Total Environ.* **2006**, *364*, 175–187.

77. Mukai, H.; Furuta, N.; Fujii, T.; Ambe, Y.; Sakamoto, K.; Hashimoto, Y. Characterization of sources of lead in the urban air of Asia using ratios of stable lead isotopes. *Environ. Sci. Technol.* **1993**, *27*, 1347–1356.

78. Bollhöfer, A.; Rosman, K. Isotopic source signatures for atmospheric lead: The Northern Hemisphere. *Geochim. Cosmochim. Acta* **2001**, *65*, 1727–1740.

3

Evapotranspiration Estimates over Non-Homogeneous Mediterranean Land Cover by a Calibrated "Critical Resistance" Approach

Paolo Martano

CNR—Istituto di Scienze dell'Atmosfera e del Clima, U.O. S. Lecce, Via Monteroni, 73100 Lecce, Italy; E-Mail: p.martano@isac.cnr.it

Academic Editors: Daniele Contini and Robert W. Talbot

Abstract: An approach based on the Penman-Monteith equation was used to estimate the actual evapotranspiration from local meteorological data over non-homogeneous land cover in a Mediterranean site in the south-east of Italy, with two six month data sets from two different years of measurements (2006 and 2009). The "critical resistance" formulation was used in different forms to model the surface resistance, together with some modifications to take into account the soil moisture content. One, two, or three model parameters were estimated, one of them related to the atmospheric resistance and the others to the surface resistance, and the calibration was made by either linear regression or nonlinear minimization of a proper cost function, depending on the applicability. Two kinds of cost functions were tested, the first depending on both the latent heat flux and the difference between screen air temperature and surface radiometric temperature, and the second depending on the temperature difference only. In all cases the calculated fluxes give better results with respect to both a flux-gradient approach and a complementarity based method, that require comparable data inputs. However the calibration by the temperature differences only, that requires no turbulent flux measurements, considerably increases the statistical uncertainty of the calibration parameters. The inclusion of the soil moisture did not significantly improve the model results in the considered site.

Keywords: evapotranspiration; Penman-Monteith equation; critical resistance; non-linear regression; surface temperature

1. Introduction

Evapotranspiration is a main component of surface energy, water budget, and of the water cycle, and is of great importance, from micrometeorological to climate and water availability projection studies. Thus improving evapotranspiration estimations and measurements from the local to the global scale is still a research challenge for the meteorological and hydrological scientific community [1].

The use of radiative energy input (either total or net radiation) together with surface radiometric temperature data ("brightness temperature", [2]) as basic information for evapotranspiration and turbulent surface fluxes estimations has been studied for several years in different forms, generally in the context of fluxes recovered from satellite data [3].

Many different methods have been proposed, some of which are directly based on the correlation between the spatial variability of the surface temperature and the evaporation fluxes and need a large amount of area distributed data to span the maximum regional surface temperature interval for calibrating the 0–1 interval of evaporative fraction [4,5]. Other methods require just "local" input data, *i.e.*, representative of the flux footprint source area, and parameter values equally related to the local surface characteristics. A complex description of the surface features is used in Land Surface Models (LSM) and Soil Vegetation Atmosphere Transfer models (SVAT), with the necessity of a large number of "effective" surface parameters not always easy to be calculated and/or calibrated [6,7]. It follows that more complex models are not always associated with better performances in flux estimations [3]. Here, attention is focused on simpler local data models with a small number of parameters, that are in general based either on vertical flux-gradient relations (e.g., SEBS, [8]) that can also distinguish between bare soil and canopy contributions [9], or thermodynamic algorithms often based on the complementarity principle [10,11]. Depending on the specific model, the surface temperature data need to be integrated by additional local meteorological information (air temperature/humidity, wind speed) and/or parameters containing basic surface information such as aerodynamic roughness and Leaf-Area Index (LAI). However, in methods involving flux-gradient approaches the evaluation of the surface roughness parameters can be misleading, as they are normally associated to the aerodynamic surface temperature that is often quite different (even several degrees) from the measured radiometric temperature [12]. These problems can be overcome in principle by a direct calibration of the surface resistance parameters with respect to the air-surface measured temperature differences and fluxes. In addition, direct calculations of heat fluxes from temperature gradients are restricted by the availability and the uncertainty of surface temperature measurements. Indeed the statistical error due to the uncertainty of the measured air-surface temperature difference can be of some tens in percent value, and can correspond to an error of the same amount or greater in the calculated fluxes [13]. In these cases the evaporation flux is generally obtained as residue of the surface energy balance, carrying even larger uncertainty. On the other hand, local data based methods for directly calculating the latent flux from Priestley-Taylor (PT) or Penman equations have also been proposed, often in association with the complementarity principle [10]. Venturini *et al.* [11] used the complementarity principle together with the PT equation and a thermodynamic calculation based on a two point interpolation of the water vapour saturation curve to obtain the surface water vapour pressure, estimating the evapotranspiration flux with no need of local parameter calibration in principle. However they did not exclude the possibility of a better local tuning of the interpolation procedure to ensure good applicability in different surface conditions.

The Penman-Monteith (PM) equation [14] has been widely used for evaluating reference evapotranspiration over homogeneous canopies (crops), where it has already been assumed as reference procedure to determine crop water requirements [15]. In the case of reference evapotranspiration the standardized characteristics of the surface (canopy height, LAI, and surface moisture) allow the use of parameters depending only on the crop type for both the surface resistance and the atmospheric resistance in the case of daily averaged evapotranspiration estimates, as discussed by Allen *et al.* [16]. In this scheme the parameter values for estimating evapotranspiration for different well watered crops have been tabulated and standardized for practical application in the calculation of the water requirements in different climate conditions [15]. The reference evapotranspiration from the PM equation has been also used to estimate the real evapotranspiration over cultivars and also natural vegetated areas by the use of correction coefficients based for example on satellite vegetation indexes [17,18]. However, even for the reference evapotranspiration, it was noted that different day/night values for the surface resistance have to be used to guarantee correct daily averages if the PM expression is applied for periods shorter than a diurnal cycle [16]. This shows a somehow expected dependence of the surface resistance response on the incident radiative fluxes.

An analysis of the surface "bulk" resistance R_0 in the PM equation was performed by Alves and Pereira [19], together with an intercomparison with the functional forms of the multiplicative functions proposed by Jarvis [20] to model a canopy resistance. They showed that the PM equation requires a well-defined functional dependence of the surface resistance from the meteorological variables of the same equation (available energy flux, air humidity deficit, and temperature) and that these functional forms are in agreement with the semi-empirical expressions adopted by Jarvis [20] for available radiative flux and humidity deficit.

A dimensional analysis of the PM equation by Katerji and Perrier [21] had shown similar results for the functional form of the surface resistance, that appeared to be a function of a "critical resistance" R^*, which can be considered the surface resistance value that decouples the canopy evapotranspiration from the atmospheric resistance effects; this with a form in agreement with the results found by Alves and Pereira [19].

Rana *et al.* [22] used the concept of critical resistance to model the actual evapotranspiration over crops. In this model the surface (crop) resistance depends linearly on the critical resistance but the linear coefficients are shown to depend on the crop moisture availability. They found the measured predawn leaf water potential of the crop to be a good parameter for describing how the canopy moisture availability affects the surface resistance and obtained a simple model for the actual evapotranspiration even in water stress conditions for homogeneous crops, where the predawn water potential can be determined by direct measurements. Rana *et al.* [23] also simplified this result proposing a "crop coefficient" model in which an empirical relation between the leaf water potential and the measured soil water moisture is used in the field. Katerji and Rana [24] also showed that the "critical resistance" approach to model a canopy resistance that varies with local meteorology gives better estimations of the actual evapotranspiration compared with a constant locally determined canopy resistance, especially for hourly scale measurements and tall crops, in different kinds of sites and cultivars.

The above mentioned studies show that the PM equation admits a dependence of the surface resistance on the local meteorological variables in the same equation (energy flux input, air humidity deficit, and temperature) and that its functional form is somehow independent of the surface characteristics but defined by similarity properties of the equation as function of a "critical resistance". They also show that varying the functional coefficients with the water stress allows PM based models to be used for actual evapotranspiration estimates also in semi-arid climate conditions. These considerations about the prescribed functional form of the surface resistance in the PM equation also suggest the possibility of extending its application to natural canopies, in which however both the atmospheric resistance parameters and the surface resistance parameters can be difficult to be determined, either from the canopy characteristics or by direct measurements, and should be recovered by calibration procedures.

Stewart [25] applied the PM equation with the Jarvis [20] approach for the surface resistance to model the evapotranspiration of a pine forest, after experimentally calibrating several model parameters by a multivariate method. He obtained the best results with a surface resistance depending non-linearly on the local meteorology and the available soil moisture, for which an explained variance of the order of 70% was obtained. Shuttleworth and Wallace (SW) [26] proposed a model based on a development of the PM equation to take into account partially shaded canopy surfaces (sparse crops). It separates the radiation and evapotranspiration contributions from the canopy and the partially shaded underlying surface using a LAI-based parameterization. Stannard [27] compared the results of the SW model with measurements in a semiarid region and with the results of the PT and PM model, again using the Jarvis [20] formulation for the canopy resistance. He found that the SW model, that separates the soil from the canopy evaporative contributions, performs better that the PM big-leaf model with a single surface resistance in the site, characterized by sparse vegetation. The enhanced soil evaporation appeared to be important just after the rain events but in general the surface resistance was quite insensitive to soil moisture variations in the considered site. The previous approaches, together with use of the Jarvis [20] canopy resistance formulation, require several parameters to be calibrated with respect to locally measured fluxes, as well as information about the turbulent surface transfer coming from external parameterizations or turbulence measurements.

In the present work the PM equation is applied to calculate the actual evapotranspiration over non homogeneous mixed Mediterranean vegetation. The "critical resistance" is used to model the surface resistance with the advantage of a much reduced number of unknown parameters, testing different formulations including a parallel combination of a soil and a canopy resistance to take into account the non-homogeneous vegetation cover (Figure 1). All the parameters characterizing the atmospheric and the surface resistance are calculated by linear or non-linear regression methods, depending on the applicability, using data of measured surface radiometric temperature, either together with the measured surface fluxes or not. The intention here is of exploring the performance of a PM based model with few calibrated parameters and data inputs over non-homogeneous Mediterranean land cover, with the possible use of the PM equation in a method for the surface heat fluxes partitioning that uses the surface temperature solely for calibration purposes.

2. Experimental and Modelling Section

2.1. The Modelling Scheme

The modelled latent heat flux Q_{em} is expressed in the PM formulation as [14]:

$$Q_{em} = [sE + \rho C_p q_{sat} (1-H_r)/R_a][s + \gamma(1 + R_0/R_a)]^{-1} \qquad (1)$$

where H_r is the air relative humidity , q_{sat} the saturation specific humidity, E the available energy flux, s the slope of the saturation curve (from the Clapeyron formula), γ the psychrometric constant, C_p and ρ the specific heat and the density of the air. All variables in the right hand side (r.h.s.) are measured with slow response sensors of common use in automated meteorological stations, with the exception of the surface "bulk" resistance R_0 and the aerodynamic resistance R_a.

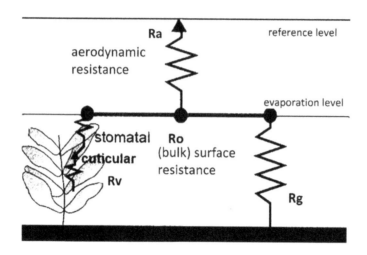

Figure 1. Schematic view of the resistance arrangements. R_v and R_g are the canopy and the ground resistance respectively (refer to text).

A standard way to parameterize the aerodynamic resistance R_a by means of the wind speed U is to use a constant drag coefficient C_h depending on the surface roughness characteristics and on the measurement height in the form [14]:

$$R_a = (C_hU)^{-1} = (P_1U)^{-1} \tag{2}$$

Here the scalar transfer coefficient C_h is considered to be the same as that for sensible heat transfer, and in the following paragraphs P_1, P_2, P_3 are the unknown parameters to be calibrated by regression methods.

In non-homogeneous land cover the surface "bulk" resistance R_0 can be considered as a function of a canopy resistance R_v and a soil surface (ground) resistance R_g (Figure 1).

Over vegetated surfaces the form of the canopy resistance R_v has been discussed by Katerji and Perrier [21] who show by dimensional reasoning that it depends on the meteorological variables through a critical resistance R^*, defined as the resistance value decoupling the evapotranspiration flux from the aerodynamic resistance; this is found to be:

$$R^* = \rho L_e q_{sat}(1 - H_r)(1 + \gamma/s)/E \tag{3}$$

The canopy resistance is then expected to be a function of R*, say $R_v/R_a = f(R_*/R_a)$. Rana *et al.* [22] use a linear function f to model the surface evaporation on homogeneous crops using different parameter values for different water stress conditions of the cultivars.

Here it is assumed a vegetation canopy contribution R_v to the surface resistance R_0 of the form,

$$R_v/R_a = C_vR^*/R_a = P_2UR^* \tag{4}$$

where C_v is an unknown coefficient that can depend also on the vegetation water availability and $P_2 = C_vC_h$. Note that with this form of the surface resistance it is possible to eliminate R_a in the PM equation and directly express C_v using the Bowen ratio Bo and the surface-air temperature difference ΔT. Indeed using the definition of R_a through the sensible heat flux Q_s,

$$Q_s = EBo/(Bo + 1) = \rho C_p \Delta T/R_a = \rho C_p \Delta T \; C_h U => \rho C_p/(ER_a) = (Bo/\Delta T)(Bo + 1)^{-1} \qquad (5)$$

and noting that R_a appears in the PM equation and in $R*/R_a$ only through the expression $\rho C_p/(ER_a)$, after some algebra it follows that:

$$C_v = [(s + q_{sat} (1 - H_r)/\Delta T) - \gamma/Bo] (Bo + 1)/[q_{sat} (1 - H_r)(1 + \gamma/s)/\Delta T] \qquad (6)$$

This directly expresses C_v by measured quantities only.

In non-vegetated soil conditions the soil resistance R_g, is a function of a "soil wetness parameter" [28] that can be a strongly nonlinear function of the soil specific water content w. The wetness parameter is easily shown to be mathematically equivalent to a surface resistance formulation, as they can be expressed as function of one another. Kondo et al. [29] expressed the wetness parameter by a "wetness function" of the soil moisture $F(w)$. They also used simple experimental equipment, in which evaporation and soil water content were monitored by weighing soil samples in two contiguous evaporation plates one of which is maintained in saturation, and $F(w)$ is determined comparing the measured saturation evaporation and actual evaporation measured from the known soil water content [29]. All the mentioned measurements are obtained by weighing the plates at proper time intervals during the drying period. Here, using their expression for the wetness parameter as a function of $F(w)$, and the mathematical equivalence between the wetness parameter and the surface resistance, after some straightforward algebraic manipulations, the soil surface resistance can be written as:

$$R_g/R_a = (C_s F(w)/D_a) (C_h U) = (C_s C(w_{sat} - w)^p/D_a) (C_h U) = P_3 U(w_{sat} - w)^p/D_a \qquad (7)$$

Following Kondo et al. [29], an empirical power function $F(w) = C(w_{sat} - w)^p$ has been used here, where p and C are regression parameters depending on the soil texture, w_{sat} is the saturation specific water content, $D_a = 0.000023(T_s/273.2)^{1.75}$ with T_s indicating the soil temperature and $P_3 = C_s C_h$. For the purpose of computing D_a in Equation (7) the soil surface temperature T_s was approximated by the radiometric surface temperature T_0 in the following paragraphs.

2.2. The Data Set and the Calibration of Parameters

The data set is obtained from the CNR ISAC web data base in Lecce [30], where a description of the experimental set up can also be found. Data were collected in the experimental base in the Salento University Campus, 5 km SW from the city of Lecce, South-Eastern Italy, in a Mediterranean landscape covered by shrubs and trees (mainly olive and pine, all evergreen with a quite constant foliage cover), with some university buildings around. The climate is typically Mediterranean, with precipitations concentrated in autumn-winter and a warm dry summer, with soil conditions that tend to be wet from December to February, and quickly drying up between May and June. A 16 m height mast is equipped with standard meteorological instruments routinely collecting half-hour averages of standard meteorological variables, including surface radiometric temperature, measured by a surface Everest 4004.GL infrared thermometer, and net radiation (measured by a Rebs Q*7.1 net radiometer). A fast response eddy correlation system (Solent-Gill ultrasonic anemometer, and Campbell Krypton Hygrometer Kh20) outputs half-hour averaged turbulent fluxes and variances in streamline coordinates, as described by Martano [31], as well as the averaged wind speed components. An ancillary Campbell meteorological station also collected half-hour averages of wind speed, temperature, humidity and soil surface data. Soil temperature, and heat flux are measured at 2 and 5 cm depth by Campbell 107L temperature sensors and Hukseflux HFP01-05 heat flux plates, and soil moisture averages between 0–5 and 30–35 cm depth by Decagon EC-20 (year 2006) and Decagon EC-5 (year 2009) sensors. These sensors were used following the constructor's calibration, however the soil moisture sensor measurements were rescaled by the local soil moisture saturation content, measured as mentioned in Section 3 [32].

Surface radiometric temperature measurements should take into account the thermal (longwave) emissivity of the local surface [2]. The Everest 4004.GL infrared thermometer has an adjustable output (directly of temperature) and a home-made calibration was performed by leaving a sample of the (dry) soil surface in an indoor temperature controlled environment for two days and then calibrating the adjustable temperature output until equating the measured radiometric surface temperature to the measured thermometric environment temperature [33] .

Two periods from the data archive were used in the simulations: January–July 2006 and January–July 2009, taking into account the quality of measurements and their availability for any soil moisture condition. However some differences must be remarked on between the two periods. In the year 2009 the experimental site was displaced about 200 m·NW from the previous position and during summer 2008 a fire destroyed a significant part of the shrub vegetation that covered the soil surface. When the base was located in the new position, the soil moisture sensors EC-20 were substituted by EC-5 type for better accuracy measurements.

Both linear and nonlinear regression methods were applied here to estimate the unknown parameters in R_0 and R_a from the measured data. In the first case Equation (6) was used to directly calculate expressions for C_v (see next section). In the second case of non-linear dependence of the parameters a cost function S was minimized by the Levenberg-Marquardt method [34,35].

To obtain information on both the surface resistances to the evaporation and the scalar transfer coefficient, the cost function S depending on both the latent heat flux and the air-surface temperature difference was written in the following form:

$$S(P_1, P_2, P_3) = (2N)^{-1} \sum_1^N ([Q_e - Q_{em}(P_1, P_2, P_3)]^2/\sigma_Q^2 + [\Delta\Theta_{v0} - \Delta\Theta_{vm}(P_1, P_2, P_3)]^2/\sigma_\Theta^2) \qquad (8)$$

Here Q_e represent the experimental latent heat flux data, and $\Delta\Theta_{v0}$ is the measured surface-atmosphere virtual potential temperature difference between the surface radiometric temperature T_0 and the air temperature T_a (depending on the considered level of measurement) from a data set of N measurements. The modelled temperature difference $\Delta\Theta_{vm}(P_1, P_2, P_3)$ can be expressed using the energy budget closure and the drag laws as:

$$\Delta\Theta_{vm} = [E - Q_e]/(C_p\rho UP_1) \qquad (9)$$

The modelled latent heat flux $Q_{em}(P1, P_2, P_3)$ can be a function either of some or all of the three parameters, depending on the functional forms actually chosen to represent the surface resistance, that are functions of the model parameters P_1, P_2 and P_3 as in Equations (2), (4) and (7) respectively.

Finally, an attempt was made to obtain a nonlinear regression for the unknown P_1, P_2, P_3 without fluxes measurements, just putting together the information of $Q_{em}(P_1, P_2, P_3)$ and $\Delta\Theta_{vm}$ in the following cost function,

$$S(P_1,P_2,P_3) = N^{-1}\sum_1^N([\Delta\Theta_{v0} - \Delta\Theta_{vm}(P_1,P_2,P_3)]^2/\sigma\Theta^2) \qquad (10)$$

where now the temperature difference $\Delta\Theta_{vm}$ is expressed by the modelled latent heat flux Q_{em}:

$$\Delta\Theta_{vm}(P_1,P_2,P_3) = [E - Q_{em}(P_1,P_2,P_3)]/(C_p\rho UP_1) \qquad (11)$$

Note that the surface temperature appears only for parameter calibration purposes, thus, once the parameters have been obtained, the fluxes calculations are actually not limited by the surface temperature data availability (e.g., satellite periodicity for remote sensed data).

The expected statistical errors σ_Q and σ_T for Q_e and Θ_v were chosen as 30 $W \cdot m^{-2}$ and 3 K respectively. The first value is suggested by an evaluation of the experimental errors and is in general agreement with an expected uncertainty coming from the energy budget estimation of the latent heat flux in the same site [36]. The second is derived by an estimation of the standard deviation between the local radiometric surface temperature T_0 and the soil near-surface temperature T_s. It is also comparable with the dispersion obtained by comparison with some satellite derived land surface temperature data obtained by the National Oceanic and Atmospheric Administration Polar Operational Environmental Satellites Advanced Very High Resolution Radiometer (NOAA POES AVHRR) over the same site [37], as shown in Figure 2 (see also below). Indeed, this choice for the statistical uncertainties is in agreement with the expected obtained convergence values for S that should be of the order of unity [35].

Note that the evaluation of a (constant) value for $\sigma\Theta$ is not strictly necessary when Equation (10) is used instead of Equation (8), as a (constant) multiplier does not affect the position of the minimum of S, that is the values of P_1, P_2, P_3.

The NOAA POES AVHRR data were also used to give an approximate correction of the surface infrared thermometer measurements, for the effect of non-homogeneous reflectivity/emissivity of the land cover in the area affecting the measurements. This correction is shown in Figure 2 which compares the measured day/night maxima of the surface radiometric temperature T_0, that has a source area of the order of 1 m^2, and some data of maximum day/night land surface temperature T_{sat} from NOAA POES AVHRR (four daily measurements), with surface resolution of 1.1 km^2, that is actually comparable with the measured turbulent fluxes footprint source area. Satellite data refer to 1 pixel containing the measurement site for the available days in the period May–June 2005. Although only maximum night/day values could be compared, a bias of about 3.5 K in the average is apparent between the two data sets in daytime (Figure 2). It is possibly due to differences in the surface effective emissivity related to the very different source areas between the local infrared temperatures and the satellite measurements, as the thermal reflectance can be strongly affected by the small scale non-homogeneities of the surface such as terrain, rocks, buildings and different types of vegetation [2]. Thus the satellite derived data were chosen as reference brightness temperature to correct local radiometric temperature measurements due both to their much larger areal resolution of about 1 km^2 that is comparable to the footprint area of the measured turbulent fluxes, and in view of their availability for practical applications.

Models based on the PM equation also assume the identity between the scalar transfer coefficient for heat and water vapor C_h, and the closure of the surface energy budget, so that the experimental heat fluxes Q_e and Q_s were obtained here by imposing the energy budget closure to the Bowen ratio Bo (from eddy correlation measurements) and the measured available energy flux E: $Q_e = E(1 + Bo)^{-1}$ and $Q_s = EBo(1 + Bo)^{-1}$.

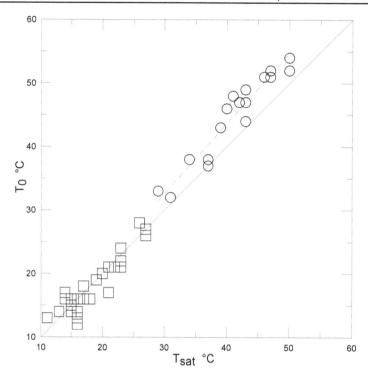

Figure 2. Plot of the surface radiometric temperature day/night maxima from the local infrared thermometer (T_0) *versus* those from the NOAA POES AVHRR satellite dataset T_{sat}, for 1 pixel containing the measurement site and available days in May–June 2005. Squares: night-time, circles: daytime. The dashed line is a regression for the daytime data only, and its average vertical distance from the x = y continuous line is 3.5 K.

3. Results and Discussion

The PM equation was used to partition the total energy flux in latent and sensible heat fluxes with different expressions for the atmospheric and the surface resistances in three cases as explained below.

CASE (1): The surface resistance R_0 is considered equal to R_v in Equation (4) and two modelling approaches were compared. In the first, used mainly for comparison purposes, P_2 of Equation (4) was calculated by just averaging Equation (6) over the data set and using Equation (5) to avoid parameterizations of R_a. In the second case R_a was parameterized by Equation (2) with $R_0/R_a = P_1UR^*$, and the non-linear minimization of Equation (8) was used to calculate the parameters $P_1 = C_h$ and $P_2 = C_vC_h$.

CASE (2): Information about the soil humidity content was introduced here.

In the first approach, used mainly for comparison purposes as in case (1), the coefficient P_2 in Equation (4) was written as $P_2 = Aw^b$, empirically assuming a power law dependence of the surface resistance coefficient on the surface soil moisture. A and b were found by linear regression of the logarithm of Equation (6) over the data set, again using Equation (5) without parameterization of R_a. In the second approach R_a was parameterized by Equation (2) and the total surface resistance was modelled as a parallel contribution of R_v and R_g from Equations (4) and (7) (Figure 1):

$$1/R_0 = 1/R_v + 1/R_g$$
$$\text{so that} \quad R_a/R_0 = (P_2UR^*)^{-1} + (P_3UC(w_{sat} - w)^P/D_a)^{-1} \tag{12}$$

The coefficients P_1, P_2, P_3 were calculated by non-linear minimization of Equation (8).

In this case it is assumed that in partially vegetated areas the bulk surface resistance R_0 can be expressed as two independent contributions of the vegetated and the non-vegetated surface fractions. Their resistances, R_v and R_g respectively, are in parallel combination, weighted with the unknown parameters P_2 and P_3.

CASE (3): The nonlinear regressions in items (1) and (2) were considered again (*i.e.*, parameterization of both R_a and R_0 with and without soil moisture contribution) but the minimization procedure was now applied to Equations (10) and (11) instead of (8) and (9). In this case no information about the surface fluxes is needed for the calibration procedure and T_0 only is used for the parameter calibration.

In items (2) and (3), when taking into account the soil moisture, some essential characteristics of the soil response should also be known. They were estimated in the same way as Kondo *et al.* [26], using an experimental set-up of two simultaneous evaporation plates and a precision balance, as mentioned in Section 2, that allow the estimation of the parameter p and the coefficient C in Equation (7), together with the saturation soil moisture w_{sat} (Figure 3).

Figure 3. The two plates experimental arrangement. The plates are 35 cm in diameter and 4 cm in depth and are placed on an insulating sheet. The inserted sensors (thermometer Campbell 107L and soil moisture sensor Decagon EC-5) are also shown, together with the precision balance.

The results are shown in Figure 4, for the experimental data and the regression function $F(w) = C(w_{sat} - w)^p$ with $w_{sat} = 0.42 \pm 0.02$ as obtained for the soil saturation moisture in the experiment. The obtained parameters are $p = 12$ and $C = 4500$. However an estimate of C is actually not strictly necessary in this application, because a regression coefficient C_s depending on the actual fractional vegetation cover in the experimental site must be always estimated in the model calibration, and C and C_s, could be both included in the regression parameter P_3 (Equation (7)).

Both the 2006 and 2009 data sets were divided into two non-overlapping subsets (alternating days), one of them was used for parameter calibration and the other for fluxes calculation with the obtained parameters. The computations are made over hourly averaged data of the turbulent fluxes and the other meteorological variables, P_1, P_2, P_3 being the only model parameters to be estimated.

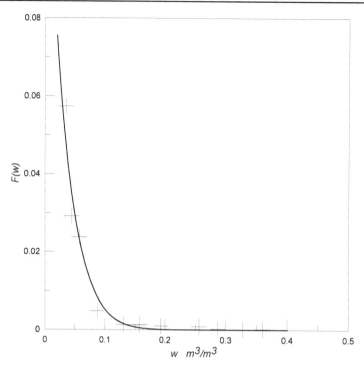

Figure 4. Plot of the wetness function $F(w)$ used in Equation (7) *versus* the soil moisture w, as obtained by the two plates evaporation experiment, with soil samples from the experimental site as mentioned in Sections 2 and 4. The continuous line shows the regression function $F(w) = C(w_{sat} - w)^p$ with $w_{sat} = 0.42$, $p = 12$ and $C = 4500$. The point sizes are of the order of their experimental uncertainty.

The results are shown in Table 1 for the obtained parameter values and the statistics of the vapor fluxes comparison, and in Figures 5–7 for the calculated fluxes for the 2006 and 2009 data sets.

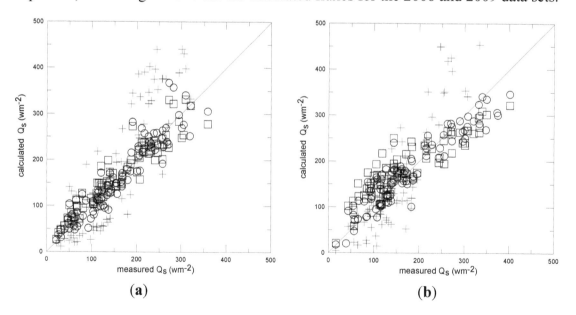

Figure 5. (**a**) Calculated *versus* measured sensible heat flux (hourly data) for year 2006 for case (1); (**b**) Calculated *versus* measured sensible heat flux for year 2009 for case (1). Plus signs: flux-gradient method, circles: single parameter linear model, squares: two parameters nonlinear model. The continuous line indicates x = y.

Figure 5a,b show the sensible heat flux Q_s calculated by the flux-gradient method, together with the sensible heat flux $Q_s = E - Q_{em}$ calculated from the surface energy budget with the latent heat flux Q_{em} obtained for case (1). They are compared with the measured sensible heat fluxes. In the flux-gradient method two constant values of C_h are used after being obtained from Equation (2) averaged over the two data sets respectively with R_a obtained as in Equation (5). They are $C_h = 10 \pm 6 \times 10^{-3}$ (2006) and $C_h = 9 \pm 7 \times 10^{-3}$ (2009).

It appears that the use of the PM equation instead of a direct application of the flux-gradient calculations reduces the scatter of the calculated fluxes that is probably due to the uncertainty on the air-surface temperature difference which can be can be quite significant. The PM equation does not make a direct use of the temperature difference to recalculate the fluxes, although this information is contained both in the linear and non-linear regressions of the parameters calibration (see Equations (6) and (8)). The required data input are almost the same for the two methods, with the addition of just the air humidity required only in the PM equation.

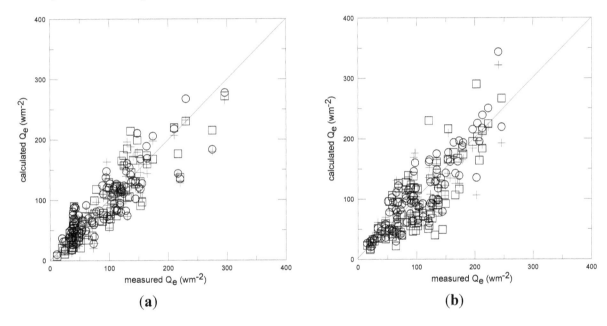

Figure 6. (a) Calculated *versus* measured latent heat flux (hourly data) for year 2006 for case (2); (b) Calculated *versus* measured latent heat flux for year 2009 for case (2). Circles: two parameters log-linear model, squares: three parameters nonlinear model, plus signs: single parameter linear model of case (1) for comparison. The continuous line indicates x = y.

Figure 6a,b show the calculated latent heat fluxes for case (2) in which the additional information about the soil surface moisture is introduced, together with the results for the first item of case (1) for comparison. Taking into account the soil surface moisture content does not apparently give relevant improvements in the flux estimates. This should imply that the relevant part of the evapotranspiration is related to the permanent vegetation cover in the surrounding areas (evergreen trees) and not to the evaporation from the soil surface. Indeed in the measurement site the soil surface usually exhibits a quick drying up in a few days even after a strong rain event, but leaving a reasonable moisture quantity at few tens of centimeters underground even for very long dry periods. In this way it works as a cover that limits soil evaporation while preserving underground moisture in the root zone, and thus maintaining plant evapotranspiration at the same time. Similar results about the model sensitivity to the soil moisture were found by Stannard [27], also in a semiarid site. The details of the statistical analysis of the regressions are reported in Table 1, where a calculation of the fractional evaporative contribution from the vegetated surface *(fev)* in the resistance parallel model is also shown *(fev = <R₀/Rᵥ>*, with *<...>* indicating the average over the data set).

Table 1. Regression parameters and statistics for the calculated *versus* the measured latent heat fluxes. *1par*: one parameter resistance model, *2par*: two parameters resistance model, *3par:* three parameters resistance model, *slope:* slope of the regression line through origin, *corr:* correlation coefficient, *sigma:* RMS dispersion of the calculated fluxes with respect to the regression line. Calculated parameters P_1, P_2, P_3 with their statistical uncertainties for the non-linear *(nl)* or linear *(lin)* regressions, calculated parameters A and p for the logarithmic linear *(lin)* regressions. Calculated fractional evapotranspiration flux from the vegetation *(fev)* for the parallel resistance in the three-parameter resistance model.

2006								
Model	*Case*	*Sigma (W·m⁻²)*	*Slope*	*corr.*	$P_1 (C_h, 10^{-3})$	$P_2 (10^{-2})$	$P_3 (10^2)$ $A(10^{-2})$	p	*fev*
1par, lin	*(1)*	27	0.94	0.93		3.2 ± 1.5			
2par, nl	*(1)*	28	0.99	0.97	8 ± 0.4	3 ± 0.4			
2par, lin	*(2)*	27	0.94	0.93		0.9		-0.57	
3par, nl	*(2)*	25	0.95	0.94	8.5 ± 0.4	3.8 ± 0.2	170 ± 300		$0.99 \pm .01$
2par, nl	*(3)*	25	0.85	0.94	10 ± 3	4 ± 3			
3par, nl	*(3)*	29	0.82	0.90	9 ± 3	5.1 ± 5.5	32 ± 36		0.84 ± 0.28

2009								
Model	*Case*	*Sigma (W·m⁻²)*	*slope*	*Corr.*	$P_1 (C_h, 10^{-3})$	$P_2 (10^{-2})$	$P_3 (10^2)$ $A(10^{-2})$	p	*fev*
1par, lin	*(1)*	33	0.94	0.93		3.4 ± 2			
2par, nl	*(1)*	38	0.97	0.95	6.3 ± 0.4	2.9 ± 0.1			
2par, lin	*(2)*	30	1.0	0.94				-0.31	
3par, nl	*(2)*	37	0.92	0.89	6.1 ± 0.2	4.0 ± 0.3	2.4 ± 0.5		0.89 ± 0.2
2par, nl	*(3)*	33	0.73	0.87	7.5 ± 2	4 ± 3			
3par, nl	*(3)*	39	0.88	0.88	7.2 ± 1.4	5.4 ± 3.2	1.5 ± 0.9		0.84 ± 0.2

In Figure 7a,b the results of the calculated latent heat flux for case 3 are reported. Here no information from measured fluxes is used for calibrations and for this reason the results were also compared with those coming from the application of the method by Venturini *et al.* [11], that requires no on-site calibration by the surface fluxes and has almost the same input parameters, including the surface brightness temperature, with the exception of the wind speed (and the soil surface moisture when taken into account).

In this case the PM calibrated method gives a smaller bias with respect to Venturini's method. Table 1 also shows that the scatter and the correlation are worse but still comparable with those of case (1) and (2) (flux calibration), but with a much larger uncertainty for the obtained parameters P_1 and P_2, that affects the slope of the regression curve. Table 1 also shows that:

- The parameterization of C_h (non-linear regression) adds a little more scatter with respect to the case without parameterization (linear regression), however the effect is small.
- There are some differences in the parameter P_1 between the years 2006 and 2009. This is somehow expected as the resistance model does not contain explicit parameters describing the surface/canopy morphology (e.g., LAI, roughness parameters), thus calibrations are expected to change in time with changes in canopy cover and soil use. In summer 2008 a fire destroyed part of the vegetation cover in the area, and the station was displaced about 200 meters away. The values of parameter P_1, (P_1, $= C_h$, that depends on roughness and also stability conditions [14]) show variations from 2006 to 2009 that are larger than their uncertainties in case (1) and (2).
- The addition of the soil moisture information gives no improvement in 2006 and very little improvement in 2009. This is in agreement with the calculated fractional evaporative contribution by the vegetation (*fev*, Table 1) and the consequently small bare soil fractional evaporative contribution (*1-fev*), and with the small sensitivity (large uncertainties) for the calculated parameter P_3. Besides changes in the surface cover, the small differences in the effect of the soil moisture between year 2006 and 2009 could be due to differences in precipitation and soil conditions, with about 650 mm annual precipitation and an average aridity index of about 0.35 in 2006, and almost 800 mm precipitation and an (anomalous) aridity index of almost 0.7 in 2009, in the measurement site [38].
- The elimination of the measured fluxes in the calibration procedure (case 3) sensibly increases the statistical uncertainty in the parameter P_1 and P_2, and this means a smaller sensitivity in the parameter evaluation when the model is calibrated using the surface-air temperature differences only.

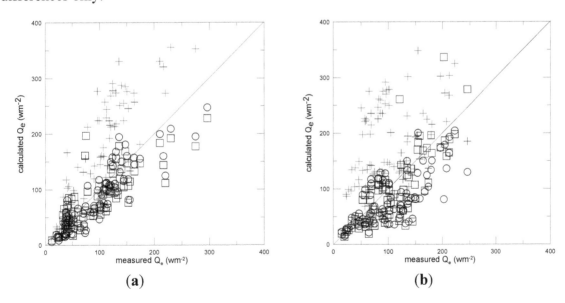

Figure 7. (a) Calculated *versus* measured latent heat flux (hourly data) for year 2006 for case (3); **(b)** Calculated *versus* measured latent heat flux for year 2009 for case (3). Circles: two parameters non-linear model, squares: three parameters nonlinear model, plus signs: Venturini's (2008) model results for comparison. The continuous line indicates x = y.

4. Summary and Conclusions

An approach based on the PM equation was used to estimate the evaporation flux over non-homogeneous terrain in a Mediterranean site in the south east of Italy for two six month data sets from the years 2006 and 2009.

Some expressions for the surface resistance, all based on the "critical resistance" approach were used in the PM equation with either one, two, or three regression parameters. The parameters were calibrated with the latent heat flux and the air-surface temperature difference data, and an attempt was also made to use the temperature differences only. The main conclusions are the following.

(1) The results compare favorably with those from a flux-gradient method that uses a value of the heat transfer coefficient C_h, calibrated on the same set of input data. In this case the use of the PM equation avoids the introduction of possibly quite large uncertainties coming from the direct use of surface-air temperature gradients.

(2) The proposed expressions for the aerodynamic and surface resistances do not contain explicit parameters describing the surface/canopy morphology (e.g., LAI, roughness parameters). As a consequence the calibrated model parameters should be used during time periods limited by the morphological surface changes, then recalibrated.

(3) The information about the surface soil moisture gave very little improvement to the results, only in 2009, a year of enhanced precipitations and anomalous surface moisture conditions, after a little displacement of the station and some differences in vegetation cover with respect to 2006. Although combined resistance models have proven to be generally more successful in modelling evapotranspiration in mixed surface areas [26,27], the surface soil moisture contribution does not seem to be essential for evapotranspiration modelling in the considered site. Here most of the evapotranspiration is likely to be due to the evergreen arboreous vegetation transfer of moisture from the root zone underground, that is often decoupled from the surface moisture. Thus parameter P_3 appears large and fairly uncertain, especially in 2006.

(4) The calculated fluxes showed less accurate but still fairly reasonable results when the calibration was made with respect to the air-surface temperature differences only, with no information about measured fluxes being used. It gives better results than a "complementarity principle" based method, that similarly requires no local flux calibration and is based on the same input data. The calibration with the local temperature appears to reduce the bias in the PM model. However, the statistical uncertainty of the parameter calibration increases considerably, indicating a significant decrease of the sensitivity of the model parameters in this case, compared with the calibration with the local fluxes, and affecting the slope of the regression curve. Thus the applicability of this procedure to other data sets in different conditions requires further investigations.

Acknowledgments

The author strongly acknowledges Cosimo Elefante and Fabio M. Grasso for their help in the management of the experimental base and the data archive. Acknowledgements are due also to Gian P. Marra for providing the satellite temperature data.

Conflicts of Interest

The author declares no conflict of interest.

References

1. Wang, K.; Dickinson, R.E. A review of global terrestrial evapotranspiration: Observation, modelling, climatology, and climatic variability. *Rev. Geophys.* **2012**, *50*, 1–54.

2. Norman, J.M.; Becker, F. Terminology in thermal infrared remote sensing of natural surfaces. *Agric. For. Meteorol.* **1995**, *77*, 153–166.

3. Kalma, J.D.; McVicar, T.M.; McCabe, M.F. Estimating land surface evaporation: A review of methods using remotely sensed surface temperature data. *Surv. Geophys.* **2008**, *29*, 421–469.

4. Bastiaanssen, W.G.M.; Menenti, M.; Feddes, R.A.; Holsltag, A.A.M. A remote sensing surface energy balance algorithm for land. I. Formulation. *J. Hydrol. (Amst.)* **1998**, *212/213*, 198–212.

5. Roerink, G.J.; Su, Z.; Menenti, M. S-SEBI a simple remote sensing algorithm to estimate the surface energy balance. *Phys. Chem. Earth Part. B Hydrol. Oceans Atmos.* **2000**, *25*, 147–157.

6. Noilhan, J.; Planton, S. A simple parameterization of land surface processes for meteorological models. *Mon. Weather Rev. (Amst.)* **1988**, *117*, 536–549.

7. Coudert, B.; Otttlé, C.; Boudevillain, B.; Demarty, J.; Guillevic, P. Contribution of thermal infrared remoste sensing data in multiobjective calibration in a dual source SVAT model. *J. Hydrometeorol.* **2006**, *7*, 404–420.

8. Su, Z. The Surface Energy Balance System (SEBS) for estimation of turbulent heat fluxes. *Hydrol. Earth Syst. Sci.* **2002**, *6*, 85–99.

9. Norman, J.M.; Kustas, W.P.; Humes, K.S. A two-source approach for estimating soil and vegetation energy fluxes from observations of directional radiometric surface temperature. *Agric. For. Meteorol.* **1995**, *77*, 263–293.

10. Granger, R.J. A complementarity relationship approach for evaporation from non-saturated surfaces. *J. Hydrol. (Amst.)* **1989**, *111*, 31–38.

11. Venturini, V.; Islam, S.; Rodriguez, L. Estimation of evaporative fraction and evapotranspiration from MODIS products using a complementary based model. *Remote Sens. Eviron.* **2008**, *112*, 132–141.

12. Stewart, J.B. Turbulent surface fluxes derived from radiometric surface temperature of sparse prairie grass. *J. Geophys. Res.* **1995**, *100*, 25429–25433.

13. Cooper, J.H.; Smith, E.A. Limitations in estimating surface sensible heat fluxes from surface and satellite radiometric skin temperatures19. *J. Geophys. Res.* **1995**, *100*, 25419–25427.

14. Garratt, J.R. *The Atmospheric Boundary Layer*; Cambridge University Press: Cambridge, UK, 1992; p. 316.

15. Allen, R.G.; Pereira, L.S.; Raes, D.; Smith, M. *Crops Evapotranspiration: Guidelines for Computing Crop Water Requirements*; *FAO Irrigation and Drainage Paper 56*; FAO: Rome, Italy, 1998; p. 300.

16. Allen, R.G.; Pruitt, W.O.; Wright, J.L.; Howell, T.A.; Ventura, F.; Snyder, R.; Itenfisu, D.; Steduto, P.; Berengena, J.; Yrisarry, J.B.; *et al.* A recommendation on standardized surface resistance for hourly calculation of reference ET_0 by the FAO 56 Penmann-Monteith method. *Agric. Water Manag.* **2006**, *81*, 1–22.

17. Glenn, E.P.; Neale, C.M.U.; Hunsaker, D.J.; Nagler, P.L. Vegetation index-based crop coefficients to estimate evapotranspiration by remote sensing in agricultural and natural ecosystems. *Hydrol. Process.* **2011**, *25*, 4050–4062.

18. Nagler, P.L.; Glenn, E.P.; Nguyen, U.; Scott, R.L.; Doody, T. Estimating riparian and agricultural actual evapotranspiration by reference evapotranspiration and MODIS Enhanced Vegetation Index. *Remote Sens.* **2013**, *5*, 3849–3871.

19. Alves, I.; Pereira, L.S. Modelling surface resistance from climatic variables? *Agric. Water Manag.* **2000**, *42*, 371–385.

20. Jarvis, P.G. The interpretation of the variations in leaf water potential and stomatal conductance found in canopies in the field. *Phil. Trans. R. Soc. Lond. B* **1976**, *273*, 593–610.

21. Katerji, N.; Perrier, A. Modélisation de l'évapotranpsiration réelle ETR d'une parcelle de luzerne: Rôle d'un coefficient cultural. *Agronomie* **1983**, *3*, 513–521.

22. Rana, G.; Katerji, N.; Mastrorilli, M.; El Moujabber, M.A. Model for predicting actual evapotranspiration under water stress conditions in a Mediterranean region. *Theor. Appl. Climatol.* **1997**, *56*, 45–55.

23. Rana, G.; Katerji, N.; Mastrorilli, M. Environmental soil-plant parameters for modelling actual crop evapotranspiration under water stress conditions. *Ecol. Model.* **1997**, *101*, 363–371.

24. Katerji, N.; Rana, G. Modelling evapotranspiration of six irrigated crops under Mediterranean climate conditions. *Agric. For. Meteorol.* **2006**, *138*, 142–155.

25. Stewart, J.B. Modelling surface conductance of pine forest. *Agric. For. Meteorol.* **1988**, *43*, 19–35.

26. Shuttleworth, J.; Wallace, J.S. Evaporation from sparse crops—An energy combination theory. *Quart. J. Roy. Meteorol. Soc.* **1985**, *111*, 839–855.

27. Stannard, D.I. Comparison of Penman-Monteith, Shuttleworth-Wallace, and modified Priestley-Taylor evapotranspiration models for wildland vegetation in semiarid rangeland. *Water Resources Res.* **1993**, *29*, 1379–1392.

28. Tsengdar, J.L.; Pielke, R.A. Estimating the soil surface specific humidity. *J. Appl. Meteorol.* **1992**, *31*, 480–484.

29. Kondo, J.; Saigusa, N.; Sato, T. A parameterization of evaporation from bare soil surfaces. *J. Appl. Meteorol.* **1990**, *29*, 385–389.

30. Basesperimentale. Available online: http://www.basesperimentale.le.isac.cnr.it/ (accessed on 16 February 2015).

31. Martano, P.; Elefante, C.; Grasso, F. A database for long term atmosphere-surface transfer monitoring in Salento Peninsula (Southern Italy). *Dataset Papers Geosci.* **2013**, *2013*, 946431.

32. Decagon Devices. Available online: http://www.decagon.com/education/calibrating-ech2o-soil-moisture-sensors-13393-04-an/ (accessed on 16 February 2015).

33. Everest Interscience Inc. Available online: http://www.everestinterscience.com/products/Enviro-Therm/Enviro-Therm.htm (accessed on 16 February 2015).

34. Beck, J.V.; Arnold, K.J. *Parameter Estimation in Engineering and Science*; Wiley: New York, NY, USA, 1977.

35. Martano, P. Inverse parameter estimation of the turbulent surface layer from single-level data and surface temperature. *J. Appl. Meteorol.* **2008**, *47*, 1027–1037.

36. Cava, D.; Contini, D.; Donateo, A.; Martano, P. Analysis of short-term closure of the surface energy balance above short vegetation. *Agric. For. Meteorol.* **2008**, *148*, 82–93.

37. The World Data Center for Remote Sensing in the Atmosphere. Available online: http://wdc.dlr.de/data_products/SURFACE/land_surface_temperature.php (accessed on 16 February 2015).

38. Martano, P.; Elefante, C.; Grasso, F. Ten years surface water and energy balance from the ISAC micrometeorological station in Salento peninsula (southern Italy). *Adv. Sci. Res.* **2014**, submitted.

4

The Possible Role of Penning Ionization Processes in Planetary Atmospheres

Stefano Falcinelli [1,2,*]**, Fernando Pirani** [3] **and Franco Vecchiocattivi** [2]

[1] Department of Chemistry, Stanford University, Stanford, CA 94305, USA

[2] Department of Civil and Environmental Engineering, University of Perugia, Via G. Duranti 93, Perugia 06125, Italy; E-Mail: franco.vecchiocattivi@unipg.it

[3] Department of Chemistry, Biology and Biotechnologies, University of Perugia, Via Elce di Sotto 8, Perugia 06123, Italy; E-Mail: fernando.pirani@unipg.it

* Author to whom correspondence should be addressed; E-Mail: stefano.falcinelli@unipg.it

Academic Editor: Armin Sorooshian

Abstract: In this paper we suggest Penning ionization as an important route of formation for ionic species in upper planetary atmospheres. Our goal is to provide relevant tools to researchers working on kinetic models of atmospheric interest, in order to include Penning ionizations in their calculations as fast processes promoting reactions that cannot be neglected. Ions are extremely important for the transmission of radio and satellite signals, and they govern the chemistry of planetary ionospheres. Molecular ions have also been detected in comet tails. In this paper recent experimental results concerning production of simple ionic species of atmospheric interest are presented and discussed. Such results concern the formation of free ions in collisional ionization of H_2O, H_2S, and NH_3 induced by highly excited species (Penning ionization) as metastable noble gas atoms. The effect of Penning ionization still has not been considered in the modeling of terrestrial and extraterrestrial objects so far, even, though metastable helium is formed by radiative recombination of He^+ ions with electrons. Because helium is the second most abundant element of the universe, Penning ionization of atomic or molecular species by $He^*(2^3S_1)$ is plausibly an active route of ionization in relatively dense environments exposed to cosmic rays.

Keywords: Penning ionization; hydrogenated molecules; metastable rare gas atoms; planetary atmospheres; mass spectrometry; electron spectroscopy; crossed molecular beams

1. Introduction

In general, atomic or molecular species can be ionized by using photons of sufficient energy: in this case we have photoionization processes. Alternately, it is possible to induce ionization via collision events involving energetic and excited particles, like energetic electrons and metastable atoms or molecules. In the latter case we have the so-called Penning ionization or chemi-ionization processes. As will be discussed in the next section, we are suggesting Penning ionization as an important route of formation for ionic species in the upper planetary atmospheres. The knowledge of cross sections, rate constants, and the stereodynamics driving such reactions should be useful to provide full basic physical chemical data needed as input for calculations that model the atmospheric chemistry of Earth and other planets of the solar system, like Mars and Mercury.

Free ionic species are extremely important for the transmission of radio and satellite signals. In fact, due to the presence of high amounts of ions and free electrons, the ionosphere plays an important role in the transmission of electromagnetic waves (especially HF radio and satellite waves). Indeed, the incidence of a HF radio wave on an ionized layer can produce total reflection under appropriate conditions, contrary to what happens on atmosphere without ionization events (whose refractive index exhibits variations generally too small to produce the total reflection of a wave) [1–3]. In addition, experimental observations demonstrate a significant improvement in the propagation of these waves after thunderstorms, especially concerning high HF bands (21-24-28 MHz) [4]. This effect is not yet fully understood and is correlated to the presence of electrical discharges in the Earth's atmosphere (statistically, from thunderstorms about 100 electrical discharges occur every second) with a huge amount of energy able to ionize and excite atomic and molecular species. For more than 20 years, studies have clearly shown that, when atoms and simple molecules (for example, H, O, rare gas atoms, and N_2 molecules) are subjected to electrical discharges, they produce a large amount of species in highly excited electronic metastable states (see Table 1). These metastable atoms or molecules can induce Penning ionization by collisions with neutral atomic or molecular targets with high efficiency [5–7]. In particular, the presence in our terrestrial atmosphere of argon (the third component of air at 0.93%), and of a minor amount of neon and helium (with concentrations of about 18 and 5 ppm, respectively), makes highly probable, in these circumstances, the production of ionized species by Penning ionization induced by these rare gas metastable atoms. It has to be noted that these processes are still not fully considered in kinetic models used to describe the physical chemistry of such environments. We are confident that our research is able to clarify the possible role played by Penning ionization processes in Earth's atmosphere, contributing to the attempt to explain some unclear phenomena, for example the "sprites" formation, *i.e.*, phenomena similar to lightning, showing a big influence on radio wave propagation. They are formed in the stratosphere (between 10 and 100 km) through electrical discharges of the duration of a few tenths of a second, which are created by the electrical potential difference between the clouds and the upper atmosphere even in the absence of thunderstorms [8]. Because these phenomena happen in certain meteorological conditions and in the presence of electrical discharges, we suggest that they could be related to the formation in the atmosphere of a strong local increase in the ionic concentration due to Penning ionization processes induced by argon, neon, and helium metastable atoms that would facilitate the transmission of the radio waves by reflection.

Table 1. Some characteristics of metastable atomic and molecular species [5].

Excited Species	Excitation Energy (eV)	Lifetime (s)
$H^*(2s^2S_{1/2})$	10.1988	0.14
$He^*(2^1S_0)$	20.6158	0.0196
$He^*(2^3S_1)$	19.8196	9000
$Ne^*(^3P_0)$	16.7154	430
$Ne^*(^3P_2)$	16.6191	24.4
$Ar^*(^3P_0)$	11.7232	44.9
$Ar^*(^3P_2)$	11.5484	55.9
$Kr^*(^3P_0)$	10.5624	0.49
$Kr^*(^3P_2)$	9.9152	85.1
$Xe^*(^3P_0)$	9.4472	0.078
$Xe^*(^3P_2)$	8.3153	150
$O^*(2p^33s^5S_2)$	9.146	1.85×10^{-4}
$H_2^*(c^3\Pi_u^-)$	11.764	1.02×10^{-3}
$N_2^*(A^3\Sigma_u^+)$	6.169	1.9
$N_2^*(A^1\Pi_g)$	8.549	1.20×10^{-4}
$N_2^*(E^3\Sigma_g^+)$	11.875	1.90×10^{-4}
$CO^*(a^3\Pi)$	6.010	$\geq 3 \times 10^{-3}$

The presence of ions in the upper terrestrial atmosphere was soon recognized after the experiment of Marconi in 1901, who succeeded in transmitting radio waves from the United Kingdom to Canada. Later, the ion CH^+ was detected in space, practically at the same time as the detection of the first neutral molecules, CH and CN [9]. Ions are extremely important in the upper atmosphere of planets, where they govern the chemistry of ionospheres. Furthermore, molecular ions have also been detected in comet tails [9]. In planetary ionospheres and in space, ions are formed in various ways, the importance of which depends on the specific conditions of the considered environment. The interaction of neutral molecules with cosmic rays, UV photons, X-rays, and other phenomena such as shock waves are all relevant processes for their production [10]. In particular, the absorption of UV photons with an energy content higher than the ionization potential of the absorbing species can induce photoionization (for most species, extreme-UV or far-UV photons are necessary). Cosmic rays are also important since they are ubiquitous and carry a large energy content (up to 100 GeV).

The detailed laboratory characterization of relevant elementary processes and their relative role, leading to the formation of atomic and molecular ions, is therefore of crucial importance. In this paper recent experimental results concerning production and characterization via Penning ionization of simple ionic species of atmospheric interest are presented. The discussed data concern the formation of free ions in collisional ionization of H_2O, H_2S, and NH_3 induced by highly excited species. As mentioned above, the effect of Penning ionization has not yet been fully considered in the modeling of terrestrial and extraterrestrial objects, even though metastable helium is known to be formed by photoelectron impact and radiative recombination of He^+ ions with electrons [11,12]. In fact, it is well-known that the 1083 nm twilight airglow arises from resonant scattering of solar photons by metastable $He^*(2^3S)$ atoms in the upper thermosphere and exosphere (see, for instance [13], and references therein). Having a radiative lifetime of 2.5 h and an energy content of 19.82 eV, Penning ionization of molecules by $He^*(2^3S_1)$ (helium is the second most abundant element of the universe) is plausibly an active route of ionization in relatively dense environments exposed to cosmic rays for two main reasons: (i) recent studies have shown that the $He^*(2^3S)$ density profiles in the Earth's atmosphere reach their maximum values at about 600 km altitude, changing with the investigated solar zenith angle, and ranging between 0.6 and 1.1 atoms/cm^3 [13]; and (ii) with respect to ionization by VUV photons and cosmic rays, Penning ionization processes are characterized by much higher cross-sections [14,15]. Moreover, concerning the role of $He^*(2^3S)$ metastable atoms in the interstellar medium, in a recent paper Indriolo et al. derived new equations for the steady state abundance of He^* using its 1083 nm absorption line data observed through the diffuse clouds toward HD 183143 [11]. By their analysis the authors inferred an upper limit for the cosmic-ray ionization rate of helium of 1.2×10^{-15} s^{-1}, which is an important route of $He^*(2^3S)$ formation via cosmic-ray ionization of helium atoms, followed by radiative recombination of the ions with electrons, as mentioned before. This ionization rate value is higher than those related to the ionization of interstellar hydrogen (various models have predicted ionization rates from 10^{-18} s^{-1} up to 10^{-15} s^{-1}) in the diffuse interstellar medium, or in comparison with the ionization of other molecules such as HD and OH having estimated ionization rates of the order of 10^{-17} s^{-1} [11]. These data allow us to highlight the importance of the chemistry associated with interstellar metastable helium, and the possible use of $He^*(2^3S)$ as a widely applicable probe of the cosmic-ray ionization rates, as suggested by Indriolo et al. [11].

Penning ionization is an autoionizing process involving the collision complex between an excited species and an atomic or molecular target. The collision will yield various ion products when the ionization potential of the target particle is lower than the excitation energy of the primary metastable species. Very common species able to induce Penning ionization are the rare gas atoms excited to their first electronic level, which is a metastable state (their lifetime is long enough to allow them to survive at least several seconds); this makes them able to ionize upon collision many atomic and molecular targets (they are characterized by a high excitation energy which ranges from 8.315 eV, in the case of $Xe^*(^3P_2)$, up to 20.616 eV, in the case of $He^*(^1S_0)$). These ionization reactions were widely studied starting in 1927 by F.M. Penning [16]. They can be schematized as follows:

$$Rg^* + M \rightarrow [Rg...M]^* \rightarrow [Rg...M]^+ + e^- \rightarrow \text{ion products} \qquad (1)$$

where Rg is a rare gas atom, M is an atomic or molecular target, $[Rg...M]^*$ represents the intermediate autoionizing collisional complex, and $[Rg...M]^+$ the intermediate transient ionic species formed by the electron ejection. This is the most efficient way for the collisional system $[Rg...M]^*$ to lose its excitation energy, since the autoionization time is usually shorter than the characteristic molecular collision time at thermal energies ($\sim 10^{-12}$ s). Then, various final product ions can be formed depending on the characteristics of the involved potential energy surfaces. Some specific features make these processes very interesting from a fundamental point of view. In fact, at high interatomic separation the metastable rare gas atoms behave as alkaline metals because of their weakly bound external electron, whereas at shorter distances the role of their "open shell" core becomes prominent, looking like halogen atoms.

Therefore, their phenomenological behavior when interacting with hydrogenated molecules (H_2O, H_2S, NH_3) is very interesting for fundamental and applied science, as shown by the many discussions in the literature about the role of hydrogen and halogen bonds [17].

However, many applications of collisional autoionization to important fields, like radiation chemistry, plasma physics, and chemistry, and the development of laser sources, are also possible. In some cases, this technique is also used to induce *gentle* ionization of complex molecules.

In general, the role of metastable rare gas atoms in Earth's atmospheric reactions is of interest for a number of reasons: (i) the quantity of rare gas atoms in the atmosphere is not negligible (argon is the third component of the air, after nitrogen and oxygen molecules: $0.934\% \pm 0.001\%$ by volume compared to $0.033\% \pm 0.001\%$ by volume of CO_2 molecules); (ii) as mentioned in the previous section, recent studies have shown that the upper atmosphere of our planet is relatively rich in $He^*(2^2S_1)$ metastable atoms [18,19]; and (iii) rate constants for ionization processes induced by metastable rare gas atoms are generally larger than those of common bimolecular chemical reactions of atmospheric interest. For example, the rate constants for Penning ionization processes are of the same order of magnitude as gas phase bimolecular reactions of $O(^1D)$, $Cl(^2P)$, or $Br(^2P)$, which are known to be relevant in the modeling of atmospheric chemistry, whereas other important reactions (as for example those involving $Br+O_3$, $O+HOCl$, and $OH+HOCl$), exhibit rate constants that are at least one order of magnitude smaller [20].

Considering the planetary atmospheric compositions, we can suppose that rare gas atoms are involved in several atmospheric phenomena not only on Earth, but also on other planets of the solar system, like Mars and Mercury. In fact, argon is the third component of the Martian atmosphere at 1.6%, while helium is a relatively abundant species (about 6%) in the low density atmosphere of Mercury. As mentioned in the previous section, a basic step in the chemical evolution of planetary atmospheres and interstellar clouds (where 89% of atoms are hydrogen and 9% are helium) is the interaction of atoms and molecules with electromagnetic waves (γ and X rays, UV light) and cosmic rays. Therefore, He^* and Ar^* formation by collisional excitation with energetic target particles (like electrons, protons, or alpha particles) and the possible subsequent Penning ionization reactions could be of importance in these environments. In particular, considering the tenuous exosphere of Mercury, it is interesting to note that in 2008 the NASA spacecraft MESSENGER discovered surprisingly large amounts of water and several atomic and molecular ionic species including O^+, OH^-, H_2O^+, and H_2S^+ (NASA and the MESSENGER team will issue periodic news releases and status reports on mission activities and make them available online). Taking into account the suggested polar deposits of water ice in this planet, McClintock and Lankton in 2007 have suggested impact vaporization mechanisms [21]. We argue that in these parts of the planetary surface, the presence of He^* and its collisions can cause subsequent ion formation. In this respect, the study of the Penning ionization of water and hydrogen sulfide by metastable rare gas atoms (He^* and Ne^*, see Section 3), producing H_2O^+, OH^+, O^+, and H_2S^+, respectively, could help in explaining the possible routes of formation for these ionic species in Mercury's exosphere. The importance of these processes and, in particular, of Penning ionization phenomena of hydrogenated molecules by He^* and Ne^* metastable atoms has been widely discussed in [22–24], to which we refer for an overview of the experimental and theoretical works performed since the 1970s by several research groups, starting with the pioneering works by Čermák and Yencha [25] and by Brion and Yee [26].

We have recently studied the ionization of water molecules by Penning ionization coupled with both Mass Spectrometry (MS) and Electron Spectroscopy (PIES) techniques [22–24]. The analysis of the obtained results suggests the selective role of a quite pronounced anisotropic attraction that controls the stereodynamics in the entrance channels of the analyzed systems. In the case of He* and Ne*-H$_2$O autoionizing collisions, we were able to confirm the presence of a strong anisotropic interaction, after previous experimental and theoretical evidences by Ohno et al. [27], Haug et al. [28], Bentley [29], and Ishida [30], respectively. In particular, we rationalized our experimental findings, taking into account the critical balancing between molecular orientation effects in the intermolecular interaction field and the ionization probability. In analyzing our PIES spectra, a novel semiclassical method was proposed: (i) it assumes ionization events as mostly occurring in the vicinities of the collision turning points (where the main part of the collision time is accumulated because the relative velocity tends to zero); and (ii) the potential energy driving the system in the relevant configurations of the entrance and exit channels, employed in the spectra simulation, has been represented by the use of a semiempirical method and given a proper analytical form [24]. This procedure was able to clearly point out how different approaches between the metastable atom and water molecules promote the formation of ions in different electronic states. In particular, it provides the angular acceptance cones where the stereo-selectivity of the processes leading to the specific formation of each one of the two energetically possible electronic states of H$_2$O$^+$ final product ions [24] (the X ^{2}B$_1$ ground state and the first A ^{2}A$_1$ excited ones) mainly manifests.

In the present work we analyze our recent PIES spectra, collected in He*, Ne*-H$_2$S, and NH$_3$ collisions, by using the same procedure already applied to water molecules, in order to compare the three sets of data, obtained under the same experimental conditions (see next Section), and to identify relevant differences in the stereodynamics of Penning ionization of the three hydrogenated molecules. In particular, our recent results recorded in He* and Ne*-H$_2$S collisional experiments clearly suggest again the formation of molecular ions both in the ground and in the first excited electronic state. They are also indicating similar anisotropic behavior for the potential energy surface describing the Rg*-H$_2$S incoming channel with an attractive component less "effective" with respect to Rg*-H$_2$O (see Section 3). In the case of ammonia, anisotropic behavior is still present, with a strong attractive interaction between Rg*-NH$_3$ collisional partners (where Rg = He, Ne), as in the Rg*-H$_2$O case (see Section 3).

2. Experimental Techniques

The experimental setup, consisting of a crossed molecular beam apparatus (three vacuum chambers which are differentially pumped at a pressure of about 10^{-7} mbar) is schematically sketched in Figure 1 and described in detail in some previous works [31–33].

In such an apparatus a molecular beam of excited atoms, mostly metastable rare gas atoms, is crossing at right angles a second beam of molecules (to be ionized). Electrons produced in photo- and/or chemi-ionization reactions are extracted and analyzed by an appropriate electron energy selector, while product ions are sent into a quadrupole mass spectrometer (located below the beam's crossing volume), where they are mass analyzed and detected. A powerful improvement of our apparatus could be realized by introducing the coincidence detection technique for the related electrons and ions coming out of the same autoionizing [Rg…M]* collisional complex (see Equation (1)).

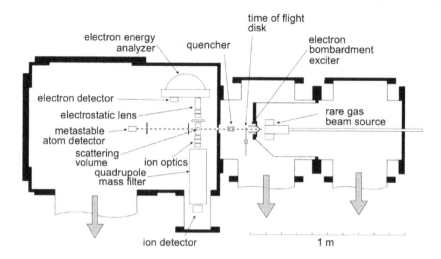

Figure 1. The crossed molecular beam apparatus used in Penning ionization studies.

The metastable rare gas beam, produced by electron bombardment or by a microwave discharge beam source, crosses at a right angle an effusive beam of H_2O, H_2S, or NH_3 target molecules. The kinetic energy of emitted electrons is recorded by the use of a hemispheric electrostatic analyzer, which is located above the crossing beam volume. In such a way we can perform, in the same experimental conditions, both photoionization (PES) and Penning Ionization Electron Spectroscopy (PIES) measurements. This can be easily achieved because the microwave discharge produces not only metastable atoms but also many He(I) and Ne(I) photons, having an energy comparable to that of metastable atoms. To calibrate the electron transmission efficiency and the electron energy scale, some He^*-H_2O, H_2S, NH_3 PIES, and Ne(I)-H_2O, H_2S, NH_3 PES spectra have been measured by exploiting the microwave discharge beam source: in the first case we use for the energy calibration the weak electron signal by He(I) photons of 21.22 eV, whereas in the latter case we record a PES spectrum, using the discharge source in pure neon producing a high intensity of $Ne(I_{\alpha,\beta})$ photons [22,34], essentially of 16.84 and 16.66 eV, respectively, in a α:β ratio of about 5.3, checked by photoelectron spectrometric measurements of Kr atoms [35,36]. By comparison of our measurements with the expected spectrum as obtained from the He(I) radiation by Kimura *et al.* [37], we calibrate the ionization energy scale. The resolution of our electron spectrometer is ~45 meV at a transmission energy of 3 eV, as determined by measuring the photoelectron spectra of Ar, O_2, and N_2 by He(I) radiation with the procedure described in previous papers [18,36]. Spurious effects due to the geomagnetic field have been reduced to ≤ 20 mG by a μ-metal shielding. Finally, the metastable atom velocity can be analyzed by a time-of-flight technique: the beam is pulsed by a rotating slotted disk and the metastable atoms are counted, using a multiscaler, as a function of the delay time from the beam opening. The secondary molecular beam can be prepared in two ways: (i) by effusion of H_2O, H_2S, or NH_3 from a glass microcapillary array kept at room temperature, or (ii) by using a supersonic source whose nozzle temperature can be changed between 300 and 700 K. The collision energy resolution is largely defined by the thermal spread of target molecules effusing from the microcapillary source (about 30% full with a half-maximum (FWHM) velocity spread at the lowest relative velocity of 1100 m/s, and about 10% FWHM for the highest probed relative velocity of 1800 m/s), and in our experiment the average collision energy is about 55 meV. Absolute values of the total ionization cross-section for different species can be obtained by the measurement of relative ion intensities in the same conditions of metastable atom and target gas density in the crossing region. This allows the various systems to be put on a relative scale, which can be normalized by reference to a known cross-section such as, in the present case, the Ne^*–Ar absolute total ionization cross-section by West *et al.* [38]. By analyzing the kinetic energy of the emitted electrons, we can perform a real spectroscopy of the collisional intermediate transient species for such a process, obtaining detailed information about the intermolecular potentials controlling the microscopic collision dynamics, the preferential geometries of approach between the collision partners, and the molecular orbitals involved in the autoionization event. On the other hand, looking at the product ions, we can determine the cross-sections for each open ionization channel, obtaining complementary information that enables us to achieve a complete understanding of the reaction dynamics for the system under study.

3. Results and Discussion

In this section recent experimental results of relevance for the characterization of simple ionic species of atmospheric interest, produced by Penning ionization of hydrogenated molecules (H_2O, H_2S, and NH_3) with He^* and Ne^* metastable atoms, are presented and discussed in a comparative way. First of all, following the observations made on the importance of Penning ionization of water by $He^*(2^3S_1, 2^1S_0)$ and $Ne^*(^3P_{2,0})$ in the understanding of the chemical composition of Mercury's atmosphere, we report here the mass spectrometric determination of the channel branching ratios for both systems as obtained at an averaged collision energy of 70 meV in the case of He^* and of 55 meV for Ne^* experiments:

$$He^* + H_2O \rightarrow He + H_2O^+ + e^- \qquad (77.3\%) \qquad (2)$$

$$\rightarrow He + H + OH^+ + e^- \qquad (18.9\%) \qquad (3)$$

$$\rightarrow He + H_2 + O^+ + e^- \qquad (3.8\%) \qquad (4)$$

$$Ne^* + H_2O \rightarrow Ne + H_2O^+ + e^- \qquad (96.2\%) \qquad (5)$$

$$\rightarrow Ne + H + OH^+ + e^- \qquad (3.2\%) \qquad (6)$$

$$\rightarrow Ne + H_2 + O^+ + e^- \qquad (0.6\%) \qquad (7)$$

No evidence has been recorded for the HeH^+, NeH^+, and HeH_2O^+, NeH_2O^+ ion production, at least under our experimental conditions. In the case of Penning ionization of ammonia by collisions with neon metastable atoms, the measured yield of possible product ions at an averaged collision energy of 55 meV is:

$$Ne^* + NH_3 \rightarrow Ne + NH_3^+ + e^- \qquad (58.8\%) \qquad (8)$$

$$\rightarrow Ne + NH_2^+ + e^- \qquad (41.2\%), \qquad (9)$$

while, for Ne^*-H_2S autoionizing collision in a thermal energy range (0.02–0.25 eV), our spectrometric measurements showed the formation of only the parent H_2S^+ product ion.

The ionization of hydrogenated molecules by metastable helium and neon atoms strongly depends on how the molecule is oriented with respect to the approaching excited rare gas atom, determining the electronic state of the intermediate ionic complex [Rg...M]$^+$ and therefore its following reactivity. In fact, for example, a direct probe of the anisotropy of the potential energy surface for Ne*-H$_2$O system can be provided by a comparative analysis of the features of electron energy spectra measured under the same conditions both in photoionization (see Figure 2—upper panel) and in Penning ionization (see Figure 2—lower panel) experiments. All the electron spectra reported in Figure 2 appeared to be composed by two bands: one (at higher electron kinetic energies) for the formation of H$_2$O$^+$ ions (in case of photoionization measurements) and of the [Ne...H$_2$O]$^+$ intermediate ionic complex (in Penning ionization determinations) in its ground X($^2B_1$) electronic state, and another one (at lower electron energy) for the formation of H$_2$O$^+$ and [Ne...H$_2$O]$^+$ in their first excited A($^2A_1$) electronic state. The formation of these two electronic states corresponds to the removal of one electron from one of the two non-bonding orbitals of the neutral molecule. The ground state X($^2B_1$) H$_2$O$^+$ ion is formed by the removal of one electron from the 2pπ lone pair orbital, while for the formation of the excited A($^2A_1$) H$_2$O$^+$ ion, the electron must be removed from the 3a$_1$ sp^2 lone pair orbital, as can be seen in Figure 3. Moreover, the "exchange mechanism" for Penning ionization suggests that one outer shell electron of the molecule to be ionized is transferred into the inner shell hole of the excited atom, and then the excess of energy produces the ejection of one electron from the collision complex [14,15].

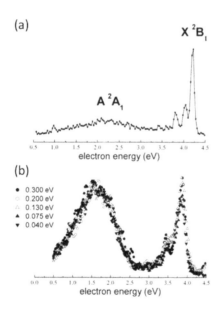

Figure 2. The obtained (**a**) PES (upper panel) and (**b**) PIES (lower panel) spectra in photoionization and Penning ionization studies of water molecules using Ne(I) photons and Ne*(^{3}P$_{2,0}$) metastable atoms, respectively. The PIES spectra reported in the lower panel are recorded at five different collision energies (0.300, 0.200, 0.130, 0.075, and 0.040 eV), by varying the nozzle temperature (between 300 and 700 K) of the supersonic source and by using, besides pure neon, two different Ne:H$_2$ mixtures (50:50 and 20:80 ratios). By looking at the relative position of the corresponding peak maxima in the PES and PIES spectra, it is possible to evaluate the evident negative energy shifts (−140 meV and −300 meV for X^{2}B$_1$ and A^{2}A$_1$ electronic states of H$_2$O$^+$, respectively) that depend on the binding energy of the autoionizing collision complex (see the text).

Therefore, we can assert that the two bands in the spectra of Figures 2 and 3 must be related to ionization events occurring with two different geometries of the system: the $X(^2B_1)$ band originates from a collision complex with the metastable neon atom perpendicular to the plane of the molecule, while for the $A(^2A_1)$ band, at the instant of the ionization, the metastable atom approaches water in the direction of the C_{2v} symmetry axis. In a recent paper [24], by analyzing the energy spectra of the emitted electrons reported in Figure 2 using a novel semiclassical method based on: (i) the assumption that the ionization events are mostly occurring in the vicinities of the collision turning points, and (ii) the formulation of the potential energy driving the system in the relevant configurations of the entrance and exit channels by the use of a semiempirical method developed in our laboratory, we were able to show clearly how different collisional approaches of the metastable atom to the water molecule lead to ions in different electronic states. In particular, we provided the angular acceptance cones where the selectivity of the process leading to the specific formation of each of the two energetically possible ionic product states of H_2O^+ emerges, showing how the ground state ion is formed when neon metastable atoms approach water mainly perpendicularly to the molecular plane, while the first excited electronic state is formed when the approach occurs preferentially along the C_{2v} axis, on the oxygen side [24].

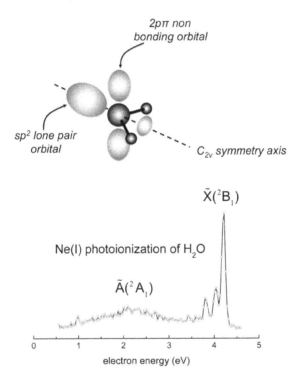

Figure 3. The structure of water molecules in the gas phase, with a C_{2v} molecular axis and a sp^2 hybridization, is depicted. The ground state $X(^2B_1)$ H_2O^+ ion is formed by the removal of one electron from the $2p\pi$ lone pair orbital, while for the formation of the excited $A(^2A_1)$ H_2O^+ ion, the electron must be removed from the $3a_1$ sp^2 lone pair orbital. The electron energy spectrum obtained in photoionization of water molecules by using Ne(I) photons is also shown. The two bands in the spectrum are related to the formation of H_2O^+ in its ground $X(^2B_1)$ electronic state (at higher electron energy), and in the first excited $A(^2A_1)$ electronic state (at lower electron energy).

By analyzing the PES and PIES spectra of Figure 2, recorded in Ne(I) and Ne^*-H_2O autoionizing collisions, respectively, we are able to obtain important findings such as the so-called "energy shifts" for the analyzed systems (see Table 2). It is well known that these observables depend on the strength, range, and anisotropy of the effective intermolecular interaction controlling the stereodynamics in the entrance Rg^*-M (neutral) and the exit Rg-M^+ (ionic) channels of ionizing collisional events. In particular, when the position of the maximum of a certain peak in the PIES spectrum exhibits a sizeable negative shift, here referred to as ε_{max}, with respect to the nominal energy, ε_0 (defined as the difference between the excitation energy of the metastable He^* or Ne^* atoms and the ionization potential of the relevant molecular state), this is an indication that a phenomenological global attractive interaction drives the collision dynamics of the partners on a multidimensional potential energy surface, including entrance and exit channels [23,36]. These energy shifts are also determined by looking at the relative position of the corresponding peak maxima in the PES and PIES spectra. For example, analyzing in a comparative way the electron spectra of Figure 2, it is possible to clearly note that both peak maxima for X and A states of produced H_2O^+ ions are shifted at lower electron kinetic energy values in the case of Penning ionization spectra with respect to the photoionization ones. The absolute values of these shifts are much higher than the difference between the energy content of the Ne^* metastable atoms and the energy of the Ne(I) photons, because of the strong attractive interaction potential between the incoming collisional partners Ne^*-H_2O (for a detailed description, see for instance [22]). In general, as the Rg^*–M attractive interaction is stronger, a more negative energy shift is expected between PES- and PIES-recorded spectra. These considerations are confirmed by our mass spectrometric determinations, and, in particular, by the total ionization cross-section measurement as a function of the collision energy reported in Figure 4. It is interesting to note that the cross-section shows a decreasing trend as the collision energy increases and, further, its absolute value is quite big (of about 50 $Å^2$ at 70 meV). Accordingly to the well-known peculiarity of Penning ionization processes [14,15], both these characteristics are a clear confirmation of the strong binding behavior of the Ne^*-H_2O autoionizing collision complex, due to the strong attractive nature of the system because of the intense molecular dipole of the water molecule ($\mu = 1.85$ D).

Analyzing the PES and PIES spectra for Ne(I) and Ne^*-H_2S, NH_3 systems, respectively, by using the same procedure already described for water molecules, we have been able to determine the related energy shifts, as shown in Figure 5 (see also Table 2). All the measured shifts have a negative value, indicating again an attractive behavior for the potential energy surface in the entrance Ne^*–M channel (where M = H_2S, NH_3) as in the case of Ne^*-H_2O autoionizing collisions. This attractive behavior is stronger in the formation of Ne^*-H_2O, whose energy shifts are -140 ± 20 and -300 ± 20 meV for the $X(^2B_1)$ and $A(^2A_1)$ bands, respectively; it is less intense in the case of Ne^*-H_2S collisions (energy shifts of -80 ± 20 and -120 ± 20 for the same two $X(^2B_1)$ and $A(^2A_1)$ bands, respectively); while in the case of Ne^*-NH_3 (energy shifts of -290 ± 20 and -140 ± 20 for the two $X(^2A_2'')$ and $A(^2E)$ bands, respectively) the attractive interaction component, affecting the entrance channel of the potential energy surface, favors those collisions producing the $[Ne...NH_3]^+$ intermediate ionic complex in its $X(^2A_2'')$—$3a_1$ ground electronic state. The formation of $[Ne...NH_3(X\ ^2A_2'')]^+$ occurs preferentially when the neon metastable atom is approaching ammonia molecules towards the direction of the $3a_1$ non-bonding lone pair located on the nitrogen atom and removes one electron from this orbital, producing Penning ionization.

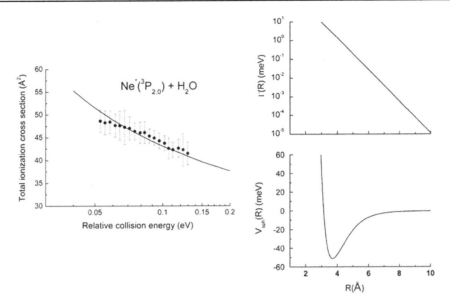

Figure 4. Left side: The total ionization cross-sections of the Ne*–H_2O autoionizing collisions as a function of the relative collision energy. The curve represents the calculation of the cross-section by applying the theory defined in terms of the so-called optical potential model, which exploits the recorded energy shifts from the electron kinetic energy spectra to probe the main characteristics of the potential energy surface for the Ne*–H_2O entrance channel (see the text and Figure 2) [14,15,22]. Right side: The spherical optical potential used for the cross-section calculation (see left side of this figure). Lower panel represents the real part of the optical potential, while the upper panel is the imaginary part (see the text and [23]).

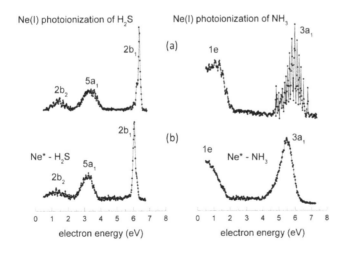

Figure 5. The obtained (**a**) PES (upper panel) and (**b**) PIES (lower panel) spectra in photoionization and Penning ionization studies of hydrogen sulfide (left side) and ammonia (right side) molecules by using Ne(I) photons and Ne*($^3P_{2,0}$) metastable atoms, respectively. By looking at the relative position of the corresponding peak maxima in the PES and PIES spectra for both investigated systems, it is possible to measure the negative energy shifts (−80 meV and −120 meV for $2b_1$ and $5a_1$ electronic states of H_2S^+, respectively; and −290 meV and −140 meV for $3a_1$ and $1e$ electronic states of NH_3^+), with the same procedure used in the Ne*-H_2O case (see Figure 2 and the text).

The measured energy shifts for the three investigated systems are compared in Table 2, where are also reported those extracted from the measured PIES spectra for He^*-H_2S and NH_3 autoionizing collisions (see Figure 6). Also in these cases we recorded negative energy shifts having a bigger value, indicating a more attractive behavior in the case of He^*–M collisions respect to Ne^*–M collisions (with M = H_2S, NH_3). This observation is in fairly good agreement with previous results obtained by Ohno *et al.* [27] and Ben Arfa *et al.* [39] for He^*–H_2O collisions, and can be justified in terms of the higher electronic polarizability of He^* atoms (118.9 and 46.9 $Å^3$ for 2^1S_0 and 2^3S_1 spin-orbit states, respectively) with respect to that of Ne^* (27.8 $Å^3$ for the $^3P_2$ state), which is the fundamental physical property of the partners, enhancing the attractive interaction behavior of the potential energy surfaces [22].

Table 2. Negative energy shifts from the PIES He^* and Ne^*-H_2O, H_2S and NH_3 spectra measured at an average collision energy of 70 and 55 meV, respectively.

Band in PIES	Shifts with Respect to Nominal Energy ε_0 (meV)	
	$H_2O^+(X\ ^2B_1) - 1b_1$	$H_2O^+(A\ ^2A_1) - 3a_1$
maximum of peak (v = 0) ε_{max} for $He^*(^3S_1)$ collisions	−440 [1]	−485 [1],*
maximum of peak (v = 0) ε_{max} for $He^*(^1S_0)$ collisions	−720 [2]	−700 [2],*
maximum of peak (v = 0) ε_{max} for Ne^* collisions	−140 ± 20	−300 ± 20 *
44% of maximum (v = 0) ε_A for Ne^* collisions	−320 ± 20	Not evaluable
Band in PIES	$H_2S^+(X\ ^2B_1) - 2b_1$	$H_2S^+(A\ ^2A_1) - 5a_1$
maximum of peak (v = 0) ε_{max} for $He^*(^3S_1)$ collisions	−150 ± 20	−190 ± 20 *
maximum of peak (v = 0) ε_{max} for $He^*(^1S_0)$ collisions	−220 ± 20	−230 ± 20 *
maximum of peak (v = 0) ε_{max} for Ne^* collisions	−80 ± 20	−120 ± 20 *
44% of maximum (v = 0) ε_A for Ne^* collisions	−195 ± 20	Not evaluable
Band in PIES	$NH_3^+(X\ ^2A_2'') - 3a_1$	$NH_3^+(A\ ^2E) - 1e$
maximum of peak (v = 0) ε_{max} for $He^*(^3S_1)$ collisions	−450 ± 20	−350 ± 20 *
maximum of peak (v = 0) ε_{max} for $He^*(^1S_0)$ collisions	−585 ± 20	Not evaluable
maximum of peak (v = 0) ε_{max} for Ne^* collisions	−290 ± 20	−140 ± 20 *
44% of maximum (v = 0) ε_A for Ne^* collisions	−840 ± 20	Not evaluable

*: evaluated from the maximum of band; [1]: from [19]; [2]: from [32].

As already discussed in previous papers [22,23], by a simple analysis of obtained PIES spectra, in the case of H_2O^+-$1b_1$, H_2S^+-$2b_1$, and NH_3^+-$3a_1$ ground states it is possible to extract not only the negative shift, ε_{max}, from the change in position of the maximum of peak but also the shift between the electron energy value where the peak intensity drops down to 44% of its maximum at the lower energy side, ε_A, with respect to the nominal energy, ε_0 [40]. For a detailed description of the theory of Penning ionization with a full discussion of the relation between the measured energy shifts from recorded PIES spectra and the main characteristics of the related potential energy surfaces, we refer to the important and basic paper of H.W. Miller published in 1970 [40]. The obtained values amount to -320 ± 20 meV for Ne^*-H_2O collisions, -195 ± 20 meV for Ne^*-H_2S collisions, and -840 ± 20 meV for Ne^*-NH_3 collisions, and they are also reported in Table 2. These findings confirm the occurrence of a very strong attractive behavior of Ne^*-NH_3 interaction promoting the Penning process, which leads to the formation of the molecular ion in the ground electronic state. A still quite strong attractive behavior drives the Ne^*-H_2O collisions, while the corresponding attraction in the Ne^*-H_2S collisions appears to be about 40% smaller than in the case of water. This consideration comes from the suggestion by Haug $et\ al.$ [28], who noted that the energy shift ε_A can be used as a proper indication of the depth of the attractive potential well of the interaction between the colliding particles in the entrance channel. This suggestion is reasonable considering the weaker attraction in the exit channel due to the very low polarizability of neon atoms $(0.40\ Å^3)$ in the ground electronic state with respect to that in the excited ones $(27.8\ Å^3)$. In our case, the estimate of ε_A can be done only for the interaction of Ne^* approaching the H_2O, H_2S, and NH_3 molecules and producing $[Ne...H_2O]^+$, $[Ne...H_2S]^+$ and $[Ne...NH_3]^+$ intermediate ionic complexes in the $(X\ ^2B_1)$-$1b_1$, $(X\ ^2B_1)$-$2b_1$, and $(X\ ^2A_2{}')$-$2b_1$ ground electronic states, respectively. An analogous evaluation of ε_A for the $(A\ ^2A_1)$ and $(A\ ^2E)$ excited states of $[Ne...H_2O]^+$, $[Ne...H_2S]^+$ and $[Ne...NH_3]^+$ intermediate ionic complexes is not possible because in these cases the recorded broad bands in the PIES spectra do not satisfy the features required by the theory [28,40–42].

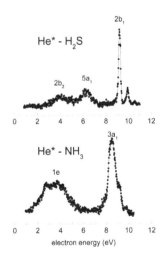

Figure 6. The obtained PIES spectra in Penning ionization studies of hydrogen sulfide (upper panel) and ammonia (lower panel) molecules by using $He^*(^3S_1, {}^1S_0,)$ metastable atoms. In the $2b_1$ band of the He^*-H_2S spectrum, the two peaks recorded at an electron energy value of 9.18 and 9.94 eV, and related to $He^*(^3S_1)$ and $He^*(^1S_0)$ collisional ionization events, respectively, appear well separated (see upper panel). The two $He^*(^3S_1, {}^1S_0,)$ metastable components are still present, even if less separated, in the wide $3a_1$ band of the He^*-NH_3 spectrum (see lower panel).

4. Conclusions

Considering that a basic step in the chemical evolution of planetary atmospheres and interstellar clouds (where 89% of atoms are hydrogen and 9% are helium) is the interaction of atoms and molecules with electromagnetic waves (γ and X rays, UV light) and cosmic rays [43], the formation of excited metastable species like He^* and Ar^* (argon is the third component of Earth and Mars' atmosphere) by collisional excitation with energetic target particles (electrons, protons, or alpha particles, coming from the solar wind) and the possible subsequent Penning ionization reactions, could be of importance in these environments. It has to be stressed that the effect of Penning ionization has not yet been fully considered in the modeling of terrestrial and extraterrestrial objects, even though metastable excited species can be formed in such environments. In this paper recent experimental results concerning production and characterization of simple ionic species of atmospheric interest are presented and discussed. Such results concern the formation of free ions in collisional ionization of H_2O, H_2S, and NH_3 induced by highly excited species, like He^* and Ne^* metastable atoms.

The analysis of the recorded electron spectra for the three presented systems clearly shows: (i) a very strong and anisotropic attractive behavior of Ne^*–NH_3 interaction promoting the Penning process; (ii) a still quite strong attractive behavior for Ne^*–H_2O collisions; and (iii) a corresponding attractive interaction in the Ne^*–H_2S collisions, which appear to be about 40% that of water. The analyzed experimental observables, in term of energy shifts, clearly indicate that H_2O and H_2S show similar behavior when these hydrogenated molecules are involved in Penning ionization induced by He^* and Ne^* metastable atoms, and in particular: (i) the interaction potential driving the formation of A $^2A_1$ excited state of H_2O^+ and H_2S^+ product ions exhibits an effective interaction potential well depth in the entrance channel of the ionization reaction, quite deeper (about 50% more in the case of water and about 25% more in the case of hydrogen sulfide) than that involved in the production of the same final ions in their X $^2B_1$ ground state (see the relative ε_{max} values reported in the Table 2); (ii) a quite pronounced anisotropic attraction controls the stereodynamics in the entrance channels of both (Ne^*–H_2O, H_2S) analyzed systems: in the case of Ne^*–H_2O autoionizing collisions a strong anisotropic interaction is operative, while in the Ne^*–H_2S system the attractive interaction appears to be weaker than the system involving water; and (iii) the entity of this smaller attraction, characterizing Ne^*–H_2S collisions, is dependent on the approach orientation between Ne^* atoms and the hydrogenated target molecules: it is about 43% with respect to the Ne^*–H_2O case for those collisions producing H_2X^+ parent ions (where X stands for O or S atom) in their X $^2B_1$ ground electronic state (this is the case for the Ne^*–H_2X collisions towards the orthogonal direction respect to the molecular plane), while it reaches about 60% of the Ne^*–H_2O interaction when the exchange mechanism of Penning ionization can exploit Ne^* collisions along the C_{2v} molecular axis in the opposite side with respect to hydrogen atoms, with the formation of H_2X^+ final ions in their first A $^2A_1$ excited electronic state. In the case of Ne^*–NH_3 collisions, a very strong and anisotropic attractive interaction is observed, characterizing the potential energy surface of the entrance channel with a much deeper well depth in the case of those collisions producing the $[Ne...NH_3]^+$ intermediate ionic complex, leading product ion in its $X(^2A_2'')$ - $3a_1$ ground electronic state (which is an opposite situation to that observed in H_2O and H_2S systems). In this case the formation of $[Ne...NH_3(X\ ^2A_2'')]^+$ occurs preferentially when the neon metastable atom approaches ammonia molecules towards the direction of the $3a_1$ non-bonding lone pair located on the nitrogen atom, removes one electron from this orbital, and gives rise to Penning ionization. Finally, the analysis of PIES spectra recorded for He*–H2S and –NH3 collisions, confirms the attractive and anisotropic behavior of the potential energy surfaces already discussed for Ne*–M interactions (with M = H2O, H2S, and NH3) governing the entrance channel for Penning ionization of such hydrogenated molecules. A similar situation for metastable helium and metastable neon atoms interacting with N2O molecules [19] has been explained by the use of a semiempirical method that accounts for the metastable atom orbital deformation because of the permanent dipole of the molecule. We have plans to extend such a model also to the systems presented here, and to test the potential energy surfaces so obtained in a comparative way. This work, is in progress in our laboratory, will allow us to achieve a deeper understanding of the stereodynamics of the collisional autoionization processes involving He* and Ne* with the hydrogenated molecules here discussed, and of the propensity to form ions.

Acknowledgments

Stefano Falcinelli is very grateful to Richard N. Zare of the Chemistry Department of Stanford University, CA, USA, for his constant encouragement and the useful suggestions during the period between June and September 2014 when Stefano Falcinelli was working in his laboratory as a visiting scholar and was conceiving the paper here presented.

Author Contributions

Stefano Falcinelli and Franco Vecchiocattivi conceived and designed the experiment; Stefano Falcinelli performed the experiment; Fernando Pirani developed the semiempirical method for potential energy surfaces calculations; all the authors analyzed the data, participated to the discussion on the obtained results, and contributed to write the paper.

Conflicts of Interest

The authors declare no conflict of interest.

References

1. Davies, K.; Rush, C.M. Reflection of high-frequency radio waves in inhomogeneous ionospheric layers. *Radio Sci.* **1985**, *20*, 303–309.
2. Hruska, F. Electromagnetic Interference and Environment. *Int. J. Circuits Syst. Signal Process.* **2014**, *8*, 22–29.
3. Qiu, H.; Chen, L.T.; Qiu, G.P.; Zhou, C. 3D visualization of radar coverage considering electromagnetic interference. *WSEAS Trans. Signal Process.* **2014**, *10*, 460–470.
4. Egano, F. *Radio Propagation Observatory.* Available online: http://www.qsl.net/ik3xtv (accessed on 14 August 2014).
5. Hotop, H. Detection of metastable atoms and molecules. In *Atomic, Molecular, and Optical Physics: Atoms and Molecules*; Dunning, F.B., Hulet, R.G., Eds.; Academic Press, Inc.: San Diego, CA, USA, 1996; pp. 191–216.

6.   Brunetti, B.G.; Falcinelli, S.; Giaquinto, E.; Sassara, A.; Prieto-Manzanares, M.; Vecchiocattivi, F. Metastable-Idrogen-Atom scattering by crossed beams: Total cross sections for H*(2s)-Ar, Xe, and CCl₄ at thermal energies. *Phys. Rev. A* **1995**, *52*, 855–858.

7.   Falcinelli, S. Penning ionization of simple molecules and their possible role in planetary atmospheres. In *Recent Advances in Energy, Environment and Financial Planning—Mathematics and Computers in Science and Engineering Series*; Batzias, F., Mastorakis, N.E., Guarnaccia, C., Eds.; WSEAS Press: Athens, Greece, 2014; pp. 84–92.

8.   Rodger, C.J.; Nunn, D. VLF scattering from red sprites: Application of numerical modeling. *Radio Sci.* **1999**, *34*, 923–932.

9.   Larsson, M.; Geppert, W.D.; Nyman, G. Ion chemistry in space. *Rep. Prog. Phys.* **2012**, *75*, 066901.

10. Stauber, P.; Doty, S.D.; van Dishoeck, E.F.; Benz, A.O. X-ray chemistry in the envelopes around young stellar objects. *Astron. Astrophys.* **2005**, *440*, 949–966.

11. Indriolo, N.; Hobbs, L.M.; Hinkle, K.H.; McCall, B.J. Interstellar metastable helium absorption as a probe of the cosmic-ray ionization rate. *Astrophys. J.* **2009**, *703*, 2131–2137.

12. Waldrop, L.S.; Kerr, R.B.; González, S.A.; Sulzer, M.P.; Noto, J.; Kamalabadi, F. Generation of metastable helium and 1083 nm emission in the upper thermosphere. *J. Geophys. Res.* **2005**, *110*, A08304.

13. Bishop, J.; Link, R. He (2³S) densities in the upper thermosphere: Updates in modeling capabilities and comparison with midlatitude observations. *J. Geophys. Res.* **2005**, *104*, 17157–17172.

14. Siska, P.E. Molecular-beam studies of Penning ionization. *Rev. Mod. Phys.* **1993**, *65*, 337–412.

15. Brunetti, B.G.; Vecchiocattivi, F. Autoionization dynamics of collisional complexes. In *Ion Clusters*; Ng, C., Baer, T., Powis, I., Eds.; Springer: New York, NY, USA, 1993; pp. 359–445.

16. Penning, F.M. Über Ionisation durch metastabile Atome. *Naturwissenschaften. J.* **1927**, *15*, doi:10.1007/BF01505431

17. Legon, A.C. The halogen bond: An interim perspective. *Phys. Chem. Chem. Phys.* **2010**, *12*, 7736–7747.

18. Biondini, F.; Brunetti, B.G.; Candori, P.; de Angelis, F.; Falcinelli, S.; Tarantelli, F.; Teixidor, M.M.; Pirani, F.; Vecchiocattivi, F. Penning ionization of N₂O molecules by He*(2³,¹S) and Ne*(³P₂,₀) metastable atoms: A crossed beam study. *J. Chem. Phys.* **2005**, *122*, 164307:1–164307:10.

19. Biondini, F.; Brunetti, B.G.; Candori, P.; de Angelis, F.; Falcinelli, S.; Tarantelli, F.; Pirani, F.; Vecchiocattivi, F. Penning ionization of N₂O molecules by He*(2³,¹S) and Ne*(³P₂,₀) metastable atoms: Theoretical considerations about the intermolecular interactions. *J. Chem. Phys.* **2005**, *122*, 164308:1–164308:11.

20. McClintock, W.E.; Lankton, M.R. The mercury atmospheric and surface composition spectrometer for the MESSENGER mission. *J. Phys. Chem. Ref. Data* **1989**, *21*, 1125–1499.

21. Atkinson, R.; Baulch, D.L.; Cox, R.A.; Hampson, R.F.; Kerr, J.A., Jr.; Troe, J. Evaluated kinetic and photochemical data for atmospheric chemistry: Supplement IV. IUPAC Subcommittee on kinetic data evaluation for atmospheric chemistry. *Space Sci. Rev.* **2007**, *131*, 481–521.

22. Brunetti, B.G.; Candori, P.; Cappelletti, D.; Falcinelli, S.; Pirani, F.; Stranges, D.; Vecchiocattivi, F. Penning ionization electron spectroscopy of water molecules by metastable neon atoms. *Chem. Phys. Lett.* **2012**, *539–540*, 19–23.

23. Balucani, N.; Bartocci, A.; Brunetti, B.G.; Candori, P.; Falcinelli, S.; Pirani, F.; Palazzetti, F.; Vecchiocattivi, F. Collisional autoionization dynamics of Ne*($^3P_{2,0}$)-H$_2$O. *Chem. Phys. Lett.* **2012**, *546*, 34–39.

24. Brunetti, B.G.; Candori, P.; Falcinelli, S.; Pirani, F.; Vecchiocattivi, F. The stereodynamics of the Penning ionization of water by metastable neon atoms. *J. Chem. Phys.* **2013**, *139*, 164305:1–164305:8.

25. Čermák, V.; Yencha, A.J. Penning ionization electron spectroscopy of H$_2$O, D$_2$O, H$_2$S and SO$_2$. *J. Electron Spectr. Rel. Phenom.* **1977**, *11*, 67–73.

26. Brion, C.E.; Yee, D.S.C. Electron spectroscopy using excited atoms and photons IX. Penning ionization of CO$_2$, CS$_2$, COS, N$_2$O, H$_2$S, SO$_2$, and NO$_2$. *J. Electron Spectr. Rel. Phenom.* **1977**, *12*, 77–93.

27. Ohno, K.; Mutoh, H.; Harada, Y. Study of electron distributions of molecular orbitals by Penning ionization electron spectroscopy. *J. Am. Chem. Soc.* **1983**, *105*, 4555–4561.

28. Haug, B.; Morgner, H.; Staemmler, V. Experimental and theoretical study of Penning ionisation of H$_2$O by metastable Helium He (2^3S). *J. Phys. B At. Mol. Phys.* **1985**, *18*, 259–274.

29. Bentley, J. Potential energy surfaces for excited neon atoms interacting with water molecules. *J. Chem. Phys.* **1980**, *73*, 1805–1813.

30. Ishida, T. A quasi-classical trajectory calculation for the Penning ionization H$_2$O-He*(2^3S). *J. Chem. Phys.* **1996**, *105*, 1392–1401.

31. Falcinelli, S.; Candori, P.; Bettoni, M.; Pirani, F.; Vecchiocattivi, F. Penning ionization electron spectroscopy of hydrogen sulfide by metastable Helium and Neon atoms. *J. Phys. Chem. A* **2014**, *118*, 6501–6506.

32. Arfa, M.B.; Lescop, B.; Cherid, M.; Brunetti, B.; Candori, P.; Malfatti, D.; Falcinelli, S.; Vecchiocattivi, F. Ionization of ammonia molecules by collision with metastable neon atoms. *Chem. Phys. Lett.* **1999**, *308*, 71–77.

33. Brunetti, B.; Candori, P.; Falcinelli, S.; Vecchiocattivi, F.; Sassara, A.; Chergui, M. Dynamics of the Penning ionization of fullerene molecules by metastable Neon atoms. *J. Phys. Chem. A* **2000**, *14*, 5942–5945.

34. Brunetti, B.; Candori, P.; Falcinelli, S.; Kasai, T.; Ohoyama, H.; Vecchiocattivi, F. Velocity dependence of the ionization cross section of methyl chloride molecules ionized by metastable argon atoms. *Phys. Chem. Chem. Phys.* **2001**, *3*, 807–810.

35. Brunetti, B.; Candori, P.; Falcinelli, S.; Lescop, B.; Liuti, G.; Pirani, F.; Vecchiocattivi, F. Energy dependence of the Penning ionization electron spectrum of Ne*($^3P_{2,0}$) + Kr. *Eur. Phys. J. D* **2006**, *38*, 21–27.

36. Kraft, T.; Bregel, T.; Ganz, J.; Harth, K.; Ruf, M.-W.; Hotop, H. Accurate comparison of HeI, NeI photoionization and He($2^{3,1}$S), Ne($3s\ ^3P_2,\ ^3P_0$) Penning ionization of argon atoms and dimers. *Z. Phys. D* **1988**, *10*, 473–481.

37. Kimura, K.; Katsumata, S.; Achiba, Y.; Yamazaky, T.; Iwata, S. *Handbook of HeI Photoelectron Spectra of Fundamental Organic Molecules*; Japan Scientific Societies Press: Tokyo, Japan, 1981.

38. West, W.P.; Cook, T.B.; Dunning, F.B.; Rundel, R.D.; Stebbings, R.F. Chemiionization involving rare gas metastable atoms. *J. Chem. Phys.* **1975**, *63*, 1237–1242.

39. Arfa, M.B.; le Coz, G.; Sinou, G.; le Nadan, A.; Tuffin, F.; Tannous, C. Experimental study of the Penning ionization of the H_2O molecule by He^* (2^3S, 2^1S) metastable atoms. *J. Phys. B At. Mol. Opt. Phys.* **1994**, *27*, 2541–2550.

40. Miller, W.H. Theory of Penning ionization. I. Atoms. *J. Chem. Phys.* **1970**, *52*, 3563–3571.

41. Hotop, H.; Niehaus, A. Reactions of excited atoms molecules with atoms and molecules. II. Energy analysis of penning electrons. *Z. Phys. D* **1969**, *228*, 68–88.

42. Falcinelli, S.; Rosi, M.; Candori, P.; Vecchiocattivi, F.; Bartocci, A.; Lombardi, A.; Lago, N.F.; Pirani, F. Modeling the intermolecular interactions and characterization of the dynamics of collisional autoionization processes. In *ICCSA 2013, Part I, LNCS 7971 Computational Science and its Applications*; Murgante, B., Misra, S., Carlini, M., Torre, C.M., Nguyen, H.Q., Taniar, D., Apduhan, B.O., Gervasi, O., Eds.; Springer-Verlag: Berlin Heidelberg, Germany, 2013; pp. 47–56.

43. Falcinelli, S.; Rosi, M.; Candori, P.; Vecchiocattivi, F.; Farrar, J.M.; Pirani, F.; Balucani, N.; Alagia, M.; Richter, R.; Stranges, S. Kinetic energy release in molecular dications fragmentation after VUV and EUV ionization and escape from planetary atmospheres. *Planet. Space Sci.* **2014**, *99*, 149–157.

5

Characteristics of Organic and Elemental Carbon in PM$_{2.5}$ and PM$_{0.25}$ in Indoor and Outdoor Environments of a Middle School: Secondary Formation of Organic Carbon and Sources Identification

Hongmei Xu [1,2,*], Benjamin Guinot [3], Zhenxing Shen [1], Kin Fai Ho [4], Xinyi Niu [2,5], Shun Xiao [2,6], Ru-Jin Huang [2,7,8] and Junji Cao [2,9,*]

[1] Department of Environmental Science and Engineering, Xi'an Jiaotong University, Xi'an 710049, China; E-Mail: zxshen@mail.xjtu.edu.cn

[2] Key Lab of Aerosol Chemistry & Physics, Institute of Earth Environment, Chinese Academy of Sciences, Xi'an 710061, China; E-Mails: niu.xin.yi@stu.xjtu.edu.cn (X.N.); xiaoshun@ieecas.cn (S.X.)

[3] Laboratoire d'Aerologie, Observatory Midi-Pyrenees, CNRS—University of Toulouse, Toulouse 31400, France; E-Mail: benjamin.guinot@aero.obs-mip.fr

[4] School of Public Health and Primary Care, The Chinese University of Hong Kong, Hong Kong, China; E-Mail: kfho@cuhk.edu.hk

[5] School of Human Settlements and Civil Engineering, Xi'an Jiaotong University, Xi'an 710049, China

[6] Shaanxi Meteorological Bureau, Xi'an 710014, China

[7] Laboratory of Atmospheric Chemistry, Paul Scherrer Institute (PSI), 5232 Villigen, Switzerland; E-Mail: Rujin.Huang@psi.ch

[8] Centre for Climate and Air Pollution Studies, Ryan Institute, National University of Ireland Galway, Galway, Ireland

[9] Institute of Global Environmental Change, Xi'an Jiaotong University, Xi'an 710049, China

* Author to whom correspondence should be addressed;
E-Mails: xuhongmei@mail.xjtu.edu.cn (H.X.); cao@loess.llqg.ac.cn (J.C.)

Academic Editors: Guohui Li and Robert W. Talbot

Abstract: Secondary organic carbon (SOC) formation and its effects on human health require better understanding in Chinese megacities characterized by a severe particulate pollution and robust economic reform. This study investigated organic carbon (OC) and elemental carbon (EC) in $PM_{2.5}$ and $PM_{0.25}$ collected 8–20 March 2012. Samples were collected inside and outside a classroom in a middle school at Xi'an. On average, OC and EC accounted for 20%–30% of the particulate matter (PM) mass concentration. By applying the EC-tracer method, SOC's contribution to OC in both PM size fractions was demonstrated. The observed changes in SOC:OC ratios can be attributed to variations in the primary production processes, the photochemical reactions, the intensity of free radicals, and the meteorological conditions. Total carbon (TC) source apportionment by formula derivation showed that coal combustion, motor vehicle exhaust, and secondary formation were the major sources of carbonaceous aerosol. Coal combustion appeared to be the largest contributor to TC (50%), followed by motor vehicle exhaust (25%) and SOC (18%) in both size fractions.

Keywords: OC and EC; SOC formation; very fine particles (VFP); TC sources; indoor and outdoor; school; Xi'an; China

1. Introduction

Organic carbon (OC) and elemental carbon (EC) are the major components of ambient atmospheric aerosols that originate from natural and anthropogenic combustion sources. Total carbon (*i.e.*, the sum of OC and EC or carbonaceous aerosol) contribution to the mean China urban aerosol mass ranges from 20% to 50% in both fine and coarse fractions [1–15]. OC and EC have long been studied to investigate their effects on the regional climate, air quality, and visibility in China [16–18]. However, these indicators are of concern owing to their possible adverse effects on human health, as several studies have supported a positive correlation between exposure to OC or EC and the exacerbation of cardiopulmonary diseases or the increase of daily mortality [19–22].

EC is a primary product emitted from combustion processes and OC exists in two forms: (i) Primary OC (POC), which is directly emitted into the atmosphere either from fossil fuel combustion by industries, road transportation, and the residential sector or by biomass burning; and (ii) secondary OC (SOC), which is formed through atmospheric oxidation of volatile organic compounds (VOCs) and gas-to-particle conversion processes. Huang *et al.* showed that the severe haze pollution events during winters in China were generally caused by secondary aerosol formation, which contributed 30%–77% of $PM_{2.5}$ (particulate matter with aerodynamic diameters ≤ 2.5 μm) and 44%–71% of organic aerosols [23]. The estimation of the sources and concentrations of SOC is therefore crucial to understand the formation mechanisms of haze and assess its related health effects. However, there is currently no direct analysis method available to separate and quantify POC from SOC in aerosols [24].

Indirect quantitative methods can be summarized as follows:

1. The tracer methods: (i) The widely-accepted EC-tracer method uses EC to trace any primary sources and calculate the primary OC in ambient samples given the OC/EC ratio provided from primary emissions and ambient OC and EC concentration data [24]; and (ii) the secondary organic aerosol (SOA)-tracer method identifies a series of tracer compounds in SOA based on chamber experiments.

Ratios of these tracers to SOA or SOC obtained from the chamber simulation are used to estimate SOA or SOC from different precursors using literature data of measured tracers in ambient samples [25].

2. The chemical transport models: (i) Reactive chemical transport models are used for predicting the concentrations of POC and SOC. This model includes emission, dispersion, and chemical transformation of gaseous and particulate OC [26,27]; and (ii) non-reactive transport models are used to estimate POC first, then SOC is calculated at ambient monitoring sites by subtracting the model POC from the measured OC [28].

3. Joint approaches: They include approaches such as the EC-tracer method combined with radiocarbon (^{14}C) analysis [29], radiocarbon analysis with chemical mass balance (CMB) [30], and the direct estimation of SOC with input of secondary species when applying the positive matrix factorization (PMF) model [31]. Last, aerosol mass spectrometry (AMS) is an advanced approach providing high time resolution [32].

Xi'an, the largest urban area in northwest China, is located in the middle of the Central Shaanxi Plain on the Yellow River and has a population of more than eight million. The city suffers from severe local and regional air pollution problems caused by rapid Chinese economic development over the past two decades. A total of 1741 elementary and secondary schools were recorded in 2012, including 962,000 students aged from 6 to 18 [33]. On average, over the school year period (holidays excluded), students spend 30%–50% of their time at school. Children are especially vulnerable to air pollutants because their respiratory systems are not fully developed yet [34,35]. Many studies have demonstrated that elevated particulate matter (PM) concentrations in classrooms may have a significant adverse effect on children's health and performance [36,37]. However, there are only a limited number of studies concerning the chemical properties (e.g., OC and EC) of PM in school environments in China [38–40].This study therefore aims to (i) characterize the indoor and outdoor OC and EC concentrations in PM2.5 and PM0.25 (PM with aerodynamic diameters ≤ 0.25 μm) in Xi'an during the end of winter 2012 at a middle school; (ii) investigate SOC mass contribution to OC and the SOC sources; and (iii) apportion TC contributions from coal combustion, motor vehicle exhaust, secondary formation, and other primary emissions. Outputs from this study may lead to better understanding about the SOC formation mechanisms of different size particles outdoors and indoors in Chinese cities experiencing serious PM pollution and to establishing source emission control strategies for particulate matter, especially for carbonaceous aerosol.

2. Materials and Methods

2.1. Site Description

The targeted middle school is located approximately 5 km south of Xi'an city center, in a commercial and residential area. The school environment was not under the direct influence of any industrial emissions, nor traffic as the closest main road is more than 300 m away. Indoor sampling was set up in a classroom located on the second floor, about 8 m above ground level. It was occupied by the same 30 students, aged between 12 and 14, from Monday to Friday, 8:00 am to 12:00 am and 1:30 pm to 5:30 pm local time. Indoor samplers were located in the back of the classroom and their inlets were set up 1.2 m above the ground. The same instrumental package was set up outdoors, at the end of the corridor passing alongside the surveyed classroom, about 6 m away from the indoor sampling site. The location of the outdoor sampling site was defined to account for the ambient air inside the school environment and not for the air from the street. The classroom was ventilated manually using two large windows, each 1.8×1.5 m in size, located on both sides of the classroom. One side was connected to a small street and the other one to the inner corridor. Windows remained opened during school hours but were closed after school and during the weekends.

2.2. PM Sample Collection

$PM_{2.5}$ and $PM_{0.25}$ were synchronously sampled on pre-fired (780 °C, 3 h) 47 mm Whatman quartz microfiber filters (QM/A, Whatman Inc., UK) indoors and outdoors during the period 8–20 March 2012 over 24-h period from 8:00 am local time onwards. Daily $PM_{2.5}$ samples were continuously collected using mini-volume air samplers (Airmetrics, OR, USA) at a flow rate of 5 $L \cdot min^{-1}$. Daily $PM_{0.25}$ samples were collected from the last particle size stage of a Sioutas Personal Cascade Impactor (SKC Inc., Fullerton, CA, USA) operating at 9 $L \cdot min^{-1}$. Particles whose diameter is equal to or lower than 0.25 μm are hereafter either defined as $PM_{0.25}$ or very fine particles (VFP). Their collection was controlled by an electrical timer, which allowed both indoor and outdoor sampling to run for 30 min every hour to avoid overloading, that is, 12 h for every 24 h. Twenty-six indoor and 26 outdoor samples were collected in total, excluding field blanks for this study.

2.3. PM Gravimetric and Chemical Analysis

PM samples were weighed on a Sartorius ME 5-F electronic microbalance (sensitivity ±1 μg, Sartorius, Germany) before and after sampling, after equilibration for 24 h at 20–23 °C and 35%–45% of relative humidity, in order to determine mass concentrations. The absolute errors between duplicate weights were less than 0.015 mg for blank filters and 0.020 mg for samples. The exposed quartz fiber filters were stored in a refrigerator at <-4 °C prior to chemical analysis to prevent evaporation of volatile components.

OC and EC were analyzed for each sample from a 0.5 cm^2 punch using a Desert Research Institute (DRI) Model 2001 Thermal/Optical Carbon Analyzer (Atmoslytic Inc., Calabasas, CA, USA) following the IMPROVE_A (Interagency Monitoring of Protected Visual Environment) thermal/optical reflectance (TOR) protocol. The method provided data following four OC fractions (OC1, OC2, OC3, and OC4 in a non-oxidizing Helium (He) atmosphere at 140 °C, 280 °C, 480 °C, and 580 °C, respectively), one OP fraction (a pyrolyzed carbon fraction, obtained in an oxidizing atmosphere and determined when the reflected laser light reaches its original intensity) and three EC fractions (EC1, EC2m and EC3 in an oxidizing atmosphere of 2% O_2 in a balance of 98% He at 580 °C, 780 °C, and 840 °C, respectively). The IMPROVE_A protocol defines TC as OC + EC, OC as OC1 + OC2 + OC3 + OC4 + OP, and EC as EC1 + EC2 + EC3 − OP. The detailed determination procedures of OC, EC, and the QA/QC can be viewed in Chow *et al.* [41,42] and Cao *et al.* [8].

3. Results and Discussion

3.1. OC and EC Concentrations in $PM_{2.5}$ and $PM_{0.25}$

The indoor and outdoor $PM_{2.5}$ and $PM_{0.25}$ mass concentrations are summarized in Table 1. $PM_{2.5}$ and $PM_{0.25}$ indoor concentrations averaged 141.8 ± 42.5 and 53.5 ± 22.9 $μg \cdot m^{-3}$, while the outdoor concentrations averaged 167.8 ± 58.6 and 49.1 ± 17.3 $\mu g \cdot m^{-3}$, respectively. The PM$_{0.25}$/PM$_{2.5}$ mass ratios over 24 h averaged 37.8%, ranging from 25.1% to 59.3% indoors, and averaged 29.3%, ranging from 23.0% to 41.2% outdoors. PM$_{0.25}$/PM$_{2.5}$ mass ratios hence appear to be nearly 9% higher indoors than outdoors. The indoor-to-outdoor ratio (I/O) ranged from 0.7 to 1.1 (averaged 0.8) for PM$_{2.5}$, and from 0.5 to 1.6 (averaged 1.1) for PM$_{0.25}$. This suggests not only that the contribution of PM$_{0.25}$ to PM$_{2.5}$ is higher indoors, but also that there are more PM$_{0.25}$ indoors than outdoors. Specific sources of PM$_{0.25}$ mass in the classroom may originate from students and/or teachers, such as coughing, sneezing, usage of cleaning products, and smoking [43–48]. A bias may have been created by the input of PM$_{0.25}$ coming from the street via a window, while the outdoor sampling site located within the school was less affected.

Table 1. PM$_{2.5}$ and PM$_{0.25}$ mass concentrations inside the classroom and outdoors during the sampling period in Xi'an, China.

Site	PM$_{2.5}$ ($\mu g \cdot m^{-3}$)		PM$_{0.25}$ ($\mu g \cdot m^{-3}$)		PM$_{0.25}$/PM$_{2.5}$ (%)	
	Concentration [a]	N [b]	Concentration [a]	N [b]	Average	Range
Indoor	141.8 ± 42.5	13	53.5 ± 22.9	13	37.8	25.1–59.3
Outdoor	167.8 ± 58.6	13	49.1 ± 17.3	13	29.3	23.0–41.2
I/O [c]	0.8	13	1.1	13	1.3	/

[a] Values represent average ± standard deviation; [b] Number of sample; [c] Indoor to outdoor ratio.

Descriptive statistics of TC, OC, and EC concentrations in PM$_{2.5}$ and PM$_{0.25}$ in the indoor and outdoor environments of the classroom are summarized in Table 2. The average TC, OC, and EC concentrations in PM$_{2.5}$ were 30.5 ± 9.5, 22.5 ± 6.6, and 7.9 ± 3.0 $\mu g \cdot m^{-3}$ indoors, and 33.7 ± 13.8, 24.9 ± 9.8, and 8.8 ± 4.2 $\mu g \cdot m^{-3}$ outdoors, respectively. As for the VFP fraction, TC, OC, and EC were found to be 15.8 ± 4.8, 12.1 ± 3.5, and 3.6 ± 1.4 $\mu g \cdot m^{-3}$ indoors, and 14.5 ± 3.8, 11.0 ± 2.7, and 3.6 ± 1.2 $\mu g \cdot m^{-3}$ outdoors, respectively. OC concentrations were in general 2.0 ± 0.5 times higher than EC (Table 2 and Figure 1). The PM$_{0.25}$/PM$_{2.5}$ ratios were similar for TC, OC, and EC in mass (0.4 to 0.5), which were higher than the proportions in mass of PM$_{0.25}$ in PM$_{2.5}$, suggesting that TC, OC, and EC are inclined to accumulate in the smaller particles. As also shown in Table 2, for indoors, TC, OC, and EC accounted for on average 21.5%, 15.9%, and 5.6% of PM$_{2.5}$ mass loading, and 29.4%, 22.6%, and 6.8% of PM$_{0.25}$ mass loading, respectively; corresponding for outdoors, TC, OC, and EC accounted for on average 20.1%, 14.8%, and 5.3% of PM$_{2.5}$ mass loading, and 29.6%, 22.5%, and 7.3% of PM$_{0.25}$ mass loading, respectively. These also showed that TC, OC, and EC tend to concentrate in finer particle fraction (PM$_{0.25}$), consistent with previous studies [3,13,49].

Table 2. The concentrations of OC and EC in PM$_{2.5}$ and PM$_{0.25}$ inside the classroom and outdoors.

Site	TC ($\mu g \cdot m^{-3}$)			OC ($\mu g \cdot m^{-3}$)			EC ($\mu g \cdot m^{-3}$)		
	PM$_{2.5}$ [a]	PM$_{0.25}$ [a]	Ratio [b]	PM$_{2.5}$ [a]	PM$_{0.25}$ [a]	Ratio [b]	PM$_{2.5}$ [a]	PM$_{0.25}$ [a]	Ratio [b]
Indoor	30.5 ± 9.5	15.8 ± 4.8	0.5	22.5 ± 6.6	12.1 ± 3.5	0.5	7.9 ± 3.0	3.6 ± 1.4	0.5
C/PM (%) [c]	21.5	29.4	1.4	15.9	22.6	1.4	5.6	6.8	1.2
Outdoor	33.7 ± 13.8	14.5 ± 3.8	0.4	24.9 ± 9.8	11.0 ± 2.7	0.4	8.8 ± 4.2	3.6 ± 1.2	0.4
C/PM (%) [c]	20.1	29.6	1.5	14.8	22.5	1.5	5.3	7.3	1.4

[a] Values represent average ± standard deviation; [b] PM$_{0.25}$ to PM$_{2.5}$ ratio; [c] Percentage of carbonaceous aerosol concentration to PM mass concentration.

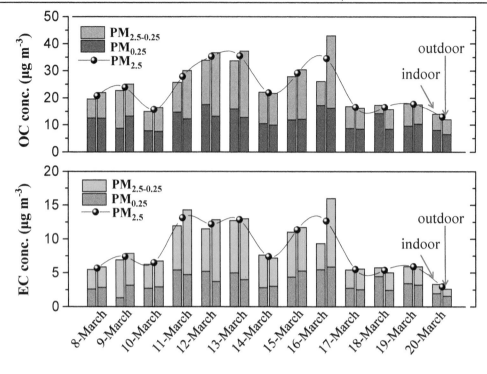

Figure 1. Time variations of OC and EC concentrations in PM$_{2.5}$ and PM$_{0.25}$ inside the classroom (left one of a couple of bars) and outdoors (right one of a couple of bars) in Xi'an, 2012.

Figure 1 shows the temporal variations of OC and EC over the study period. OC appears highly correlated with EC indoors and outdoors, in both PM$_{2.5}$ and PM$_{0.25}$ (see more details in Section 3.2.2.). OC and EC concentrations varied significantly with time, ranging in PM$_{2.5}$ from 12.0 to 43.0 µg·m^{-3} for OC, and from 2.6 to 16.0 µg·m^{-3} for EC. In PM$_{0.25}$, OC ranged from 6.5 to 16.2 µg·m^{-3} and EC from 1.6 to 5.9 µg·m^{-3}. OC and EC concentrations were slightly higher during the weekdays than weekends due to the increase in traffic volume and congestion on weekdays. However, the "weekend effect" was offset by the relatively long distance (300 m) between the sampling site and the closest road. Moreover, Xi'an has a formal heating season from 15 November to 15 March. OC and EC concentrations significantly decreased in the post-heating period of observation—typically by 40% to 50%. The highest numbers, displayed on 16 March, can be explained by the stable meteorological conditions on that day, a possible "hysteresis effect" on particle diffusion, or a delay in turning off the heating system.

3.2. Identification of SOC and Its Sources

3.2.1. Ratios of OC to EC

The OC-to-EC ratio (OC/EC) has been used to determine the emission and transformation characteristics of carbonaceous aerosols. OC/EC exceeding 2.0 indicates the presence of SOA or SOC [50]. The frequency histogram of OC/EC ratios in PM$_{2.5}$ and PM$_{0.25}$ inside the classroom and outdoors are shown in Figure 2. For each aerosol size fraction, either collected indoors or outdoors, the OC/EC values were higher than 2.0, thus suggesting the presence of SOA or SOC. OC/EC ratios were compared for PM$_{2.5}$ and PM$_{0.25}$ in Table 3. The highest ratio was recorded on 9 March, with 6.7 in PM$_{0.25}$

indoors. As shown in Figure 2, the highest frequency of OC/EC ratios was observed in the subsection of 2.5 to 3.0, with the counts of 18. The following one was between 3.0 and 3.5.

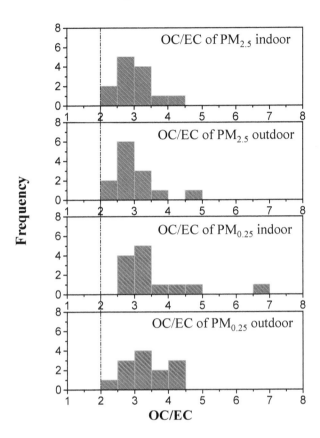

Figure 2. Frequency histogram of OC-to-EC ratios (OC/EC) in PM$_{2.5}$ and PM$_{0.25}$ inside the classroom and outdoors.

Table 3. SOC concentrations in PM$_{2.5}$ and PM$_{0.25}$ inside the classroom and outdoors, estimated from the EC-tracer method.

Site	OC/EC		(OC/EC)$_{min}$ [a]		SOC ($\mu g \cdot m^{-3}$)		SOC/OC (%)	
	PM$_{2.5}$	PM$_{0.25}$	PM$_{2.5}$	PM$_{0.25}$	PM$_{2.5}$	PM$_{0.25}$	PM$_{2.5}$	PM$_{0.25}$
Indoor	3.0	3.6	2.2	2.7	5.4	2.2	25.4	19.3
Outdoor	3.0	3.3	2.1	2.3	6.4	3.1	27.7	28.2
Average	3.0	3.5	2.1	2.5	5.9	2.6	26.6	23.8

[a] Minimum of OC to EC ratio.

Literature data suggested that average OC/EC ratios of 2.7 characterize coal combustion, with 1.1 for motor vehicle exhaust [51] and 9.0 for biomass burning [52] (Figure 3). In China, the available OC/EC ratios in PM$_{2.5}$ for cities such as Guangzhou [3] and Hong Kong [3,7] showed relatively low values (*i.e.*, 2.0 and 2.7, respectively) which suggests the predominant contribution of motor vehicle emissions. High OC/EC ratios (3.2) in the present study point to coal combustion emissions as the main contributor to winter carbonaceous aerosol levels in Xi'an.

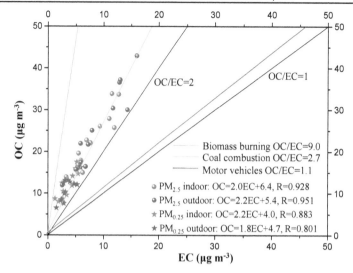

Figure 3. OC-to-EC ratios (OC/EC) in PM$_{2.5}$ and PM$_{0.25}$ inside the classroom and outdoors.

3.2.2. Relationship between OC and EC

Evidence of SOA can be inferred not only from the OC/EC ratio, but also from the absolute carbon concentration. OC and EC correlations are useful for identifying periods of significant secondary formation [24]. The regressions between OC and EC concentrations in PM$_{2.5}$ and PM$_{0.25}$ inside the classroom and outdoors are shown in Figure 3, along with data from literature [51,52]. All OC/EC ratio data points in this study are displayed between motor vehicle exhaust and biomass burning linear regressions for both PM$_{2.5}$ and PM$_{0.25}$, which reflect the combined contributions from coal combustion, motor vehicle exhaust, and biomass burning emissions. High correlation coefficients (R) of 0.93 indoors and 0.95 outdoors were found between OC and EC in PM$_{2.5}$. R values were slightly lower in PM$_{0.25}$, suggesting a clearer source signature for PM$_{2.5}$ than for PM$_{0.25}$.

According to linear regressions relating OC to EC in Figure 3, two origins of OC can be deduced—one from the intercept when EC is negligible and the other as directly correlated to EC. In the latter case, SOA formation is negligible, thus the intercept is an aggregate of non-combustion sources (e.g., biogenic OC), and can be used to represent the primary component [24]. The slopes and intercept of OC *versus* EC inside the classroom were 2.0 and 6.4 for PM$_{2.5}$ and 2.2 and 4.0 for PM$_{0.25}$, compared to those outdoors, 2.2 and 5.4 for PM$_{2.5}$ and 1.8 and 4.7 for PM$_{0.25}$ (Figure 3), implying that OC primary non-combustion emissions in PM$_{2.5}$ were higher than those in PM$_{0.25}$. The differences may be ascribed to variation of source contributions to two different particle size fractions. For example, biogenic and crustal sources generate relatively larger particles and as a result contribute more to PM$_{2.5}$ OC.

3.2.3. Estimation of SOC from the EC-Tracer Method

EC is predominantly emitted by incomplete combustion of fuels, and has stable chemical properties. Therefore, EC is a good tracer of primary anthropogenic pollutants and it has been used to estimate the concentrations of SOC, as mentioned in the introduction. The EC-tracer method is simple and straightforward, and provided that there are available measurements of OC and EC. SOC is estimated following Equation (1) [24,53]:

$$SOC = OC_{tot} - POC = OC_{tot} - EC \times (OC/EC)_{min} \tag{1}$$

where OC_{tot} is the total amount of OC in the aerosol sample and $(OC/EC)_{min}$ is the minimum ratio observed during the whole sampling period.

The fraction of SOC in OC_{tot} is shown in Table 3 and the distribution of SOC/OC (%) is illustrated in Figure 4. Estimated SOC accounted for 25.4% indoors and 27.7% outdoors of OC in $PM_{2.5}$, and 19.3% and 28.2% of OC in $PM_{0.25}$, indoors and outdoors, respectively. SOC production mainly depends on the emission rate and chemical reactivity of SOC precursors, the latter being controlled by meteorological factors such as solar radiation (given that the reactant species and their concentration are constant in the reaction) [54]. Interestingly, the SOC proportion was higher outdoors than indoors for both size fractions, which might be due to the occurrence of photochemical reactions outdoors. Generally, walls and windows reduce sunlight in buildings, and consequently photochemical formation of SOC at indoors [55,56]. This is supported in this study by the fact that the average $PM_{0.25}$ indoor SOC percentage was found to be the lowest.

The average contribution of SOC to OC in the entire dataset is comparable for $PM_{2.5}$ (26.6%) and $PM_{0.25}$ (23.8%), both indoors and outdoors (Table 3). But the attempt to discriminate indoors from outdoors leads to interesting observations. Indoors, the higher $PM_{2.5}$ SOC proportion could be due to the relatively high concentration of precursors, which can be correlated with students' perspiration and use of detergents. Outdoors, the SOC to OC ratio appeared to be similar between $PM_{2.5}$ and $PM_{0.25}$, possibly due to two distinct reasons: (i) $PM_{2.5}$ is primarily derived from direct high-temperature combustion processes emissions and complex atmospheric chemical reactions of gas-phase precursors [57,58]; and (ii) most of $PM_{0.25}$ is generated from gas-to-particle transformation reactions in the atmosphere [59]. In addition, $PM_{0.25}$ remain suspended for longer periods and can thus be transported over longer distances in the ambient air than $PM_{2.5}$, leading to more opportunities for OC in $PM_{0.25}$ to age [12,48] and produce SOC. As an illustration, the higher average percentage of SOC/OC in our dataset was found in $PM_{0.25}$ outdoors (28.2%), that is, the size fraction in the source environment where aerosols probably experience the longest residence time and the lowest contributions from primary emissions.

Figure 4 shows the SOC-to-OC ratio (%) indoors and outdoors for the two size fractions in this study. Daily $PM_{0.25}$ SOC/OC indoors varied by a factor of 19.0, ranging from 3.1% to 59.2%, followed by a factor of 4.7 for $PM_{2.5}$ indoors. SOC/OC was affected due to household cleaning product usage. Household cleaning products are an important source of indoor air pollution [60]. VOCs emitted from cleaning agents can react rapidly with indoor ozone, resulting in formation of secondary pollutants such as reactive radicals and SOA [60]. Once the cleaning product dispersed over the classroom, SOC/OC reached a relatively high level. In most schools in China, the students clean the classroom by themselves. In this school, the task was achieved every Monday, that is, on March 12th and 19th, as provided by classroom occupancy and activity records. The usage of household cleaning products generated more $PM_{0.25}$ SOC than $PM_{2.5}$ SOC, suggesting a predominant contribution of secondary rather than primary formation aerosols [3]. SOC, POC, and EC accounted for 18.1%, 56.6%, and 25.3% of TC on average in this study, respectively. The sum of these latter numbers shows that more than three quarters of TC in northern China's urban atmosphere in March may originate from primary emissions. Previous studies [4,7] that worked out the seasonal variations of SOC showed that summer was the worst period of the year, and that southern cities may suffer generally higher levels of SOC than those in the north of China. This underlines an urgent need for better understanding about SOC sources and variations, especially in East Asia.

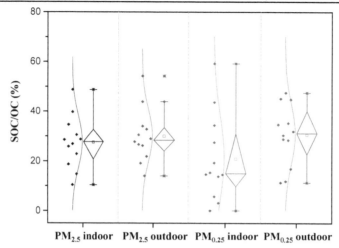

Figure 4. Box plots of SOC/OC (%) in PM2.5 and PM0.25 inside the classroom and outdoors.

3.3. Source of Carbonaceous Aerosol

3.3.1. OC and EC Relationship between Indoors and Outdoors

The correlation coefficient between indoor and outdoor data has been used as an indicator of the consistency of the source between indoors and outdoors, and the degree to which PM or its chemical compositions indoors can be attributed to infiltration from outdoors [61–65]. The relationships between indoor and outdoor OC and EC concentrations are shown in Figure 5. The coefficient (R) of determination between indoors and outdoors for OC and EC concentrations ranged between 0.70 and 0.91, suggesting a relatively high infiltration rate into the classroom, probably resulting from frequent exchanges of air.

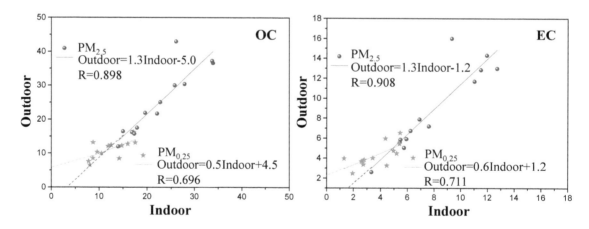

Figure 5. Indoor and outdoor comparisons of OC and EC mass concentrations in PM2.5 and PM0.25.

The indoor OC and EC concentrations in PM2.5 were strongly correlated with the corresponding outdoor concentrations, displaying R values of 0.90 and 0.91, respectively. This suggests that these carbonaceous aerosols mainly originate from an outdoor environment. As for PM0.25, fair indoor *versus* outdoor correlations of OC (R = 0.70) and EC (R = 0.71) concentrations were observed. This implies that indoor OC and EC in PM0.25 were derived, at least in part, from outdoor air. This further reveals that some indoor sources may contribute to the PM0.25 fraction, almost exclusively, such as student coughing or sneezing, usage of cleaning products in the classroom, teaching activities (e.g., usage of chalk) [39], and smoking in or nearby the classroom. Moreover, for EC, there were better correlations between indoors and outdoors for both PM$_{2.5}$ and PM$_{0.25}$ than for OC, which was consistent with results reported by Ho *et al.* [62] and Jones *et al.* [64]. Huang *et al.* [66] concluded that the majority of EC came from outdoors, mostly due to motor vehicle emissions.

3.3.2. Distribution of OC and EC Eight Fractions

Thermal carbon fractions of OC and EC are different according to the type of sources [67–69]. The fractions have been used for source apportionment analysis of carbonaceous aerosols [5,8,70,71]. Different carbon fractions in coal combustion, biomass burning (maize residue) and motor vehicle exhaust (highway with heavy traffic) source samples were studied by Cao *et al.* [5]. OC2, EC1, and EC2 can be considered as the markers of coal combustion, gasoline motor vehicle exhaust, and diesel motor vehicle exhaust, respectively [4,5,67,71].

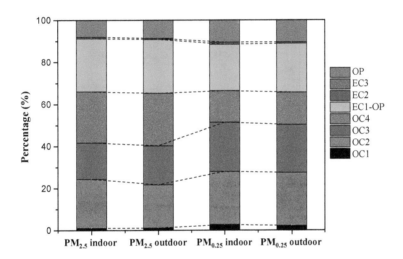

Figure 6. Percentage variations of OC and EC eight fractions in PM$_{2.5}$ and PM$_{0.25}$ inside the classroom and outdoors.

The respective contribution of the eight thermal carbon fractions to TC in PM$_{2.5}$ and PM$_{0.25}$, indoors and outdoors, are shown in Figure 6. OC2, OC3, OC4, and EC1-OP dominate. The relatively low levels of EC2 and EC3 in the samples may be due to the relatively long distance from the direct emissions of running vehicles; some motor vehicle control policy implementation may also be considered, such as the "Government Circular on Strengthening Management of the Motor Vehicle Traffic in Xi'an City." Diesel vehicles, especially trucks, were banned in Xi'an in 2007 within the second ring road from 7:00 am to 10:00 pm and important traffic regulations are in place on the second ring road during rush hour. The most abundant species—OC2, OC3, OC4, and EC1 (EC1-OP+OP)—displayed different variation trends in PM$_{2.5}$ compared to PM$_{0.25}$, but no significant differences between indoors and outdoors could be observed for the same particle size fraction. The contributions of OC2 and OC3 in TC increased from 22.0% and 17.9% in PM$_{2.5}$ to 25.3% and 23.1% in PM$_{0.25}$; meanwhile, OC4 in TC decreased from 24.6% in PM$_{2.5}$ compared to 15.2% in PM$_{0.25}$. EC1 in PM$_{2.5}$ and PM$_{0.25}$ compared to TC in the same size fractions were similar (32.6%–34.3%). These variations possibly reflected increased contributions from primary coal combustion emissions and probable secondary formation from PM$_{2.5}$ to PM$_{0.25}$. It also suggested that the contributions from gasoline-fuel motor vehicle exhaust and the corresponding secondary formation to $PM_{2.5}$ and $PM_{0.25}$ were consistent.

3.3.3. Source Apportionment of TC

A previous study in Xi'an concluded that during winter, TC was produced from gasoline exhaust (44%), coal burning (44%), biomass burning (9%), and diesel exhaust (3%) [5]. Thus, assuming in the present study that the contributions from biomass burning and diesel exhaust on EC were minimal, we considered coal combustion and motor vehicle exhaust as the main contributors. According to Cao et al. [8], their average percent contributions to EC, calculated via the ^{13}C-EC isotope mass balance approach, were 53% for coal combustion and 47% for motor vehicle exhaust. As mentioned, Watson et al. [51] reported an OC/EC ratio of 2.7 for coal combustion (CC) and 1.1 for motor vehicle exhaust (MV), which we express in Equations (2) and (3):

$$(OC/EC)_{CC} = 2.7 \tag{2}$$

$$(OC/EC)_{MV} = 1.1 \tag{3}$$

The fraction of primary OC from coal combustion (POC_{CC}) and motor vehicle exhaust (POC_{MV}) can then be expressed as Equations (4) and (5):

$$POC_{CC} = 2.7 \times EC_{CC} = 2.7 \times (53\% \times EC) = 2.7 \times 53\% \times 7.9 \ \mu g \cdot m^{-3} = 11.3 \ \mu g \cdot m^{-3} \tag{4}$$

$$POC_{MV} = 1.1 \times EC_{MV} = 1.1 \times (47\% \times EC) = 1.1 \times 47\% \times 7.9 \ \mu g \cdot m^{-3} = 4.1 \ \mu g \cdot m^{-3} \tag{5}$$

where the average EC concentration is 7.9 $\mu g \cdot m^{-3}$ in indoor $PM_{2.5}$, EC_{CC} refers to EC from coal combustion, and EC_{MV} refers to EC from motor vehicle exhaust. POC concentration can be written as the sum of POC_{CC}, POC_{MV}, and primary OC from other sources (POC_{others}). POC_{others}, containing POC from marine, soil, and biogenic emissions, can be estimated from Equation (6):

$$POC_{others} = POC - (POC_{CC} + POC_{MV}) \tag{6}$$

where the average POC concentration was 17.1 $\mu g \cdot m^{-3}$ in indoor $PM_{2.5}$. We used this information and Equation (6) to calculate POC_{others} and found POC_{others} was equal to 1.7 $\mu g \cdot m^{-3}$ in indoor $PM_{2.5}$. The respective contributions to TC of CC, MV, other primary sources, and SOC can finally be calculated from Equation (7):

$$TC = OC + EC = POC + SOC + EC = (POC_{CC} + POC_{MV} + POC_{others}) + SOC + (EC_{CC} + EC_{MV}) = (POC_{CC} + EC_{CC}) + (POC_{MV} + EC_{MV}) + POC_{others} + SOC \tag{7}$$

where SOC concentration is 5.4 $\mu g \cdot m^{-3}$ in indoor $PM_{2.5}$, ($POC_{CC} + EC_{CC}$) refers to TC from coal combustion, and ($POC_{MV} + EC_{MV}$) refers to TC from motor vehicle exhaust. As a result, the average concentrations in $PM_{2.5}$ inside the classroom attributed to coal combustion, motor vehicle exhaust, other primary sources, and SOC and calculated from Equation (7) were 15.5, 7.8, 1.7, and 5.4 $\mu g \cdot m^{-3}$, respectively.

Similarly, we estimated the relative contributions to TC of the four main sources mentioned for the three other combinations (Figure 7). On average, the results pointed out that coal combustion was the largest contributor (48.7%), followed by motor vehicle exhaust (24.4%) and SOC (18.0%), underlining that coal combustion was a dominant source of TC in Xi'an, late in the heating season for both indoors and outdoors. Moreover, coal combustion was estimated to emit more than half of the PM2.5 TC, which was 5% higher than that of PM0.25. The OC2 fraction variations reflected that SOC formation related to coal combustion emissions was elevated in PM0.25. TC originating from motor vehicle exhaust ranged from 22.9% to 25.7%, implying relatively constant inputs from this source, consistent with the variations of EC1 mentioned before. It should be noted that TC emitted from other primary sources in PM0.25 indoor was significantly higher than in other combinations. The value found for indoor PM0.25 (17.8%) suggested that there were other VFP sources existing in the classroom, which was also supported, earlier in this article, by the OC and EC relationships between indoors and outdoors.

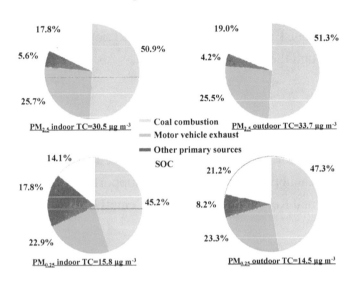

Figure 7. Relative contributions from different sources to TC in PM2.5 and PM0.25 inside the classroom and outdoors.

4. Conclusions

Daily PM2.5 and PM0.25 samples collected from 8 to 20 March 2012, inside and outside a classroom in a middle school at Xi'an, China, were used to feature OC and EC according to two different particle size fractions and to identify SOC by various methods in order to understand the SOC formation pathways and the source apportionment of TC. OC and EC constituted significant fractions of PM in atmosphere, as they accounted for 20.8% and 29.5% of PM2.5 and PM0.25 mass concentrations, respectively. High OC/EC ratios (3.2), compared to other cities in China, underlined that coal combustion emissions were the most important emission source of elevated carbonaceous aerosol levels in Xi'an. Based on the EC-tracer method, SOC accounted for 25.4% and 27.7% of OC for PM2.5 indoors and outdoors, and 19.3% and 28.2% of OC for PM0.25 indoors and outdoors, respectively. The SOC-to-OC ratios indoors and outdoors were comparable between PM2.5 and PM0.25, which is consistent with higher levels outside the classroom than inside, and might result from outdoor photochemical reactions, as well as from EC primary emissions in outdoor combustion sources. High OC and EC correlations of PM2.5 and PM0.25 indicated a similar origin of OC and EC; similarly, a decent relationship between the indoor and outdoor OC and EC concentrations suggested relatively good ventilation in the classroom, and suggested that the highest composition of indoor carbonaceous aerosols could be attributed to outdoor PM permeation.

TC source apportionment by formula derivation showed that coal combustion, motor vehicle exhaust, other primary sources, and secondary formation were the major sources of carbonaceous sources in the present study. The contribution of coal combustion to $PM_{0.25}$ displayed a lower value (46.3%) than to $PM_{2.5}$ (51.1%), while motor vehicle exhaust contribution was similar (approximately 25%) for both particle sizes. Interestingly, contributions from other primary sources increased from 5.6% in indoor $PM_{2.5}$ to 17.8% in indoor $PM_{0.25}$, pointing out the existence of special indoor VFP sources such as students coughing, sneezing, usage of cleaning products, teaching activities, smoking, and so on. Overall, carbonaceous aerosols were characterized as a major component in $PM_{2.5}$ and more importantly in $PM_{0.25}$. $PM_{0.25}$ are expected to be one of the major components influencing regional atmospheric chemistry and climate change in China. Our results therefore suggested that the importance of reducing carbonaceous aerosols in the atmosphere requires more control to enhance the policies of coal combustion and motor vehicle exhaust emissions, in addition to controlling secondary production of organic compounds from OC through their gaseous precursors.

Acknowledgments

This study was supported by different projects entitled "International (Regional) Cooperation and Exchange Projects, Research Fund for International Young Scientists of the Chinese Academy of Sciences (41150110474)," "New teachers' scientific research support plan" of Xi'an Jiaotong University (XJTU-HRT-002)," and "Fundamental Research Funds for the Central Universities (XJJ2015035)."

Author Contributions

Benjamin Guinot, Hongmei Xu, and Junji Cao designed the study. Hongmei Xu, Benjamin Guinot, and Xinyi Niu collected particulate samples. Hongmei Xu and Xinyi Niu performed the gravimetric and OC, EC analysis. Hongmei Xu analyzed the data. Hongmei Xu and Benjamin Guinot designed and wrote the paper. All authors reviewed and commented on the paper.

Conflicts of Interest

The authors declare no conflict of interest.

References

1. He, K.; Yang, F.; Ma, Y.; Zhang, Q.; Yao, X.; Chan, C.K.; Cadle, S.; Chan, T.; Mulawa, P. The characteristics of $PM_{2.5}$ in Beijing, China. *Atmos. Environ.* **2001**, *35*, 4959–4970.
2. Ye, B.M.; Ji, X.L.; Yang, H.Z.; Yao, X.H.; Chan, C.K.; Cadle, S.H.; Chan, T.; Mulaw, P.A. Concentration and chemical composition of $PM_{2.5}$ in Shanghai for a 1-year period. *Atmos. Environ.* **2003**, *37*, 499–510.
3. Cao, J.J.; Lee, S.C.; Ho, K.F.; Zhang, X.Y.; Zou, S.C.; Fung, K.; Chow, J.C.; Watson, J.G. Characteristics of carbonaceous aerosol in Pearl River Delta region, China during 2001 winter period. *Atmos. Environ.* **2003**, *37*, 1451–1460.
4. Cao, J.J.; Lee, S.C.; Ho, K.F.; Zou, S.C.; Fung, K.; Li, Y.; Watson, J.G.; Chow, J.C. Spatial and seasonal variations of atmospheric organic carbon and elemental carbon in Pearl River Delta Region, China. *Atmos. Environ.* **2004**, *38*, 4447–4456.

5. Cao, J.J.; Wu, F.; Chow, J.C.; Lee, S.C.; Li, Y.; Chen, S.W.; An, Z.S.; Fung, K.K.; Watson, J.G.; Zhu, C.S.; *et al.* Characterization and source apportionment of atmospheric organic and elemental carbon during fall and winter of 2003 in Xi'an, China. *Atmos. Chem. Phys.* **2005**, *5*, 3127–3137.

6. Cao, J.J.; Lee, S.C.; Zhang, X.Y.; Chow, J.C.; An, Z.S.; Ho, K.F.; Watson, J.G.; Fung, K.K.; Wang, Y.Q.; Shen, Z.X. Characterization of airborne carbonate over a site near Asian Dust source regions during spring 2002 and its climatic and environmental significance. *J. Geophys. Res.* **2005**, *110*, 1–8.

7. Cao, J.J.; Lee, S.C.; Chow, J.C.; Watson, J.G.; Ho, K.F.; Zhang, R.J.; Jin, Z.D.; Shen, Z.X.; Chen, G.C.; Kang, Y.M.; *et al.* Spatial and seasonal distributions of carbonaceous aerosols over China. *J. Geophys. Res.* **2007**, *112*, D22S11.

8. Cao, J.J.; Zhu, C.S.; Tie, X.X.; Geng, F.H.; Xu, H.M.; Ho, S.S.H.; Wang, G.H.; Han, Y.M.; Ho, K.F. Characteristics and sources of carbonaceous aerosols from Shanghai, China. *Atmos. Chem. Phys.* **2013**, *13*, 803–817.

9. Guinot, B.; Cachier, H.; Sciare, J.; Tong, Y.; Xin, W.; Yu, J. Beijing aerosol: Atmospheric interactions and new trends. *J. Geophys. Res. Atmos.* **2007**, *112*, doi:10.1029/2006JD008195.

10. Guinot, B.; Cachier, H.; Oikonomou, K. Geochemical perspectives from a new aerosol chemical mass closure. *Atmos. Chem. Phys.* **2007**, *7*, 1657–1670.

11. Shen, Z.X.; Cao, J.J.; Arimoto, R.; Zhang, R.J.; Jie, D.M.; Liu, S.X.; Zhu, C.S. Chemical composition and source characterization of spring aerosol over Horqinsandland in northeastern China. *J. Geophys. Res.* **2007**, *112*, doi:10.1029/2006JD007991.

12. Shen, Z.X.; Cao, J.J.; Tong, Z.; Liu, S.X.; Reddy, L.S.S.; Han, Y.M.; Zhang, T.; Zhou, J. Chemical characteristics of submicron particles in winter in Xi'an. *Aerosol Air Qual. Res.* **2009**, *9*, 80–93.

13. Shen, Z.X.; Cao, J.J.; Liu, S.X.; Zhu, C.S.; Wang, X.; Zhang, T.; Xu, H.M.; Hu, T.F. Chemical composition of PM10 and PM2.5 collected at ground level and 100 meters during a strong winter-time pollution episode in Xi'an, China. *J. Air Waste Manag. Assoc.* **2011**, *61*, 1150–1159.

14. Shen, Z.X.; Cao, J.J.; Zhang, L.M.; Liu, L.; Zhang, Q.; Li, J.J.; Han, Y.M.; Zhu, C.S.; Zhao, Z.Z.; Liu, S.X. Day-night differences and seasonal variations of chemical species in PM10 over Xi'an, northwest China. *Environ. Sci. Pollut. Res.* **2014**, *21*, 3697–3705.

15. Tao, J.; Shen, Z.X.; Zhu, C.S.; Yue, J.H.; Cao, J.J.; Liu, S.X.; Zhu, L.H.; Zhang, R.J. Seasonal variations and chemical characteristics of sub-micrometer particles (PM1) in Guangzhou, China. *Atmos. Res.* **2012**, *118*, 222–231.

16. Jacobson, M.Z. Control of fossil-fuel particulate black carbon and organic matter, possibly the most effective method of slowing global warming. *J. Geophys. Res.* **2002**, *107*, doi:10.1029/2001JD001376.

17. Menon, S. Current uncertainties in assessing aerosol effects on climate. *Annu. Rev. Environ. Resour.* **2004**, *29*, 1–30.

18. Ramana, M.V.; Ramanathan, V.; Feng, Y.; Yoon, S.-C.; Kim, S.-W.; Carmichael, G.R.; Schauer, J.J. Warming influenced by the ratio of black carbon to sulphate and the black-carbon source. *Nat. Geosci.* **2010**, *3*, 542–545.

19. Jansen, K.; Larson, T.; Koenig, J.; Mar, T.; Fields, C.; Stewart, J.; Lippmann, M. Associations between health effects and particulate matter and black carbon in subjects with respiratory disease. *Environ. Health Perspect.* **2005**, *113*, 1741–1746.

20. Lewne, M.; Plato, N.; Gustavsson, P. Exposure to particles, elemental carbon and nitrogen dioxide in workers exposed to motor exhaust. *Ann. Occup. Hyg.* **2007**, *51*, 693–701.

21. Shih, T.S.; Lai, C.H.; Hung, H.F.; Ku, S.Y.; Tsai, P.J.; Yang, T.; Liou, S.H. Elemental and organic carbon exposure in highway tollbooths: A study of Taiwanese toll station workers. *Sci. Total Environ.* **2008**, *402*, 163–170.

22. Cao, J.J.; Xu, H.M.; Xu, Q.; Chen, B.H.; Kan, H.D. Fine particulate matter constituents and cardiopulmonary mortality in a heavily polluted Chinese city. *Environ. Health Perspect.* **2012**, *120*, 373–378.

23. Huang, R.-J.; Zhang, Y.L.; Bozzetti, C.; Ho, K.-F.; Cao, J.J.; Han, Y.M.; Daellenbach, K.R.; Slowik, J.G.; Platt, S.M.; Canonaco, F.; *et al.* High secondary aerosol contribution to particulate pollution during haze events in China. *Nature* **2014**, *514*, 218–222.

24. Turpin, B.J.; Huntzicker, J.J. Identification of secondary organic aerosol episodes and quantitation of primary and secondary organic aerosol concentrations during SCAQS. *Atmos. Environ.* **1995**, *29*, 3527–3544.

25. Kleindienst, T.E.; Jaoui, M.; Lewandowski, M.; Offenberg, J.H.; Lewis, C.W.; Bhave, P.V.; Edney, E.O. Estimates of the contributions of biogenic and anthropogenic hydrocarbons to secondary organic aerosol at a southeastern US location. *Atmos. Environ.* **2007**, *41*, 8288–8300.

26. Pandis, S.N.; Harley, R.H.; Cass, G.R.; Seinfeld, J.H. Secondary organic aerosol formation and transport. *Atmos. Environ.* **1992**, *26A*, 2269–2282.

27. Strader, R.; Lurmann, F.; Pandis, S.N. Evaluation of secondary organic aerosol formation in winter. *Atmos. Environ.* **1999**, *33*, 4849–4863.

28. Hildemann, L.M.; Rogge, W.F.; Cass, G.R.; Mazurek, M.A.; Simoneit, B.R.T. Contribution of primary aerosol emissions from vegetation-derived sources to fine particle concentrations in Los Angeles. *J. Geophys. Res.* **1996**, *101*, 19541–19549.

29. Schichtel, B.A.; Malm, W.C.; Bench, G.; Fallon, S.; McDade, C.E.; Chow, J.C.; Watson, J.G. Fossil and contemporary fine particulate carbon fractions at 12 rural and urban sites in the United States. *J. Geophys. Res.* **2008**, *113*, doi:10.1029/2007JD008605.

30. Ding, X.; Zheng, M.; Edgerton, E.S.; Jansen, J.J.; Wang, X. Contemporary or fossil origin: Split of estimated secondary organic carbon in the southeastern United States. *Environ. Sci. Technol.* **2008**, *42*, 9122–9128.

31. Yuan, Z.B.; Yu, J.Z.; Lau, A.K.H.; Louie, P.K.K.; Fung, J.C.H. Application of positive matrix factorization in estimating aerosol secondary organic carbon in Hong Kong and its relationship with secondary sulfate. *Atmos. Chem. Phys.* **2006**, *6*, 25–34.

32. Sun, Y.; Zhang, Q.; Zheng, M.; Ding, X., Edgerton, E.S.; Wang, X. Characterization and source apportionment of water-soluble organic matter in atmospheric fine particles ($PM_{2.5}$) with high-resolution aerosol mass spectrometry and GC-MS. *Environ. Sci. Technol.* **2011**, *45*, 4854–4861.

33. Xi'an Municipal Bureau of Statistics and NBS Survey Office in Xi'an. Statistics Communique on the National Economy and Social Development of the City of Xi'an. In *Xi'an Statistical Yearbook*; China Statistics Press: Beijing, China, 2013; pp. 563–588.

34. Bennett, W.D.; Zeman, K.L. Deposition of fine particles in children spontaneously breathing at rest. *Inhal. Toxicol.* **1998**, *10*, 831–842.

35. Kulkarni, N.; Grigg, J. Effect of air pollution on children. *J. Paediatr. Child Health* **2008**, *18*, 238–243.

36. Mendell, M.J.; Heath, G.A. Do indoor pollutants and thermal conditions in schools influence student performance? A critical review of the literature. *Indoor Air* **2005**, *15*, 27–52.

37. Tran, D.T.; Alleman, L.Y.; Coddeville, P.; Galloo, J.-C. Elemental characterization and source identification of size resolved atmospheric particles in French classrooms. *Atmos. Environ.* **2012**, *54*, 250–259.

38. Ward, T.J.; Noonan, C.W.; Hooper, K. Results of an indoor size fractionated PM school sampling program in Libby, Montana. *Environ. Monit. Assess.* **2007**, *130*, 163–171.

39. Fromme, H.; Diemer, J.; Dietrich, S.; Cyrys, J.; Heinrich, J.; Lang, W.; Kiranoglu, M.; Twardella, D. Chemical and morphological properties of particulate matter (PM$_{10}$, PM$_{2.5}$) in school classrooms and outdoor air. *Atmos. Environ.* **2008**, *42*, 597–660.

40. Pegas, P.N.; Nunes, T.; Alves, C.A.; Silva, J.R.; Vieira, S.L.A.; Caseiro, A.; Pio, C.A. Indoor and outdoor characterization of organic and inorganic compounds in city centre and suburban elementary schools of Aveiro, Portugal. *Atmos. Environ.* **2012**, *55*, 80–89.

41. Chow, J.C.; Yu, J.Z.; Watson, J.G.; Ho, S.S.H.; Bohannan, T.L.; Hays, M.D.; Fung, K.K. The application of thermal methods for determining chemical composition of carbonaceous aerosols: A review. *J. Environ. Sci. Health* **2007**, *42*, 1521–1541.

42. Chow, J.C.; Watson, J.G.; Robles, J.; Wang, X.L.; Chen, L.-W.A.; Trimble, D.L.; Kohl, S.D.; Tropp, R.J.; Fung, K.F. Quality assurance and quality control for thermal/optical analysis of aerosol samples for organic and elemental carbon. *Anal. Bioanal. Chem.* **2011**, *401*, 3141–3152.

43. Lee, S.C.; Chang, M. Indoor and outdoor air quality investigation at schools in Hong Kong. *Chemosphere* **2000**, *41*, 109–113.

44. Lai, A.C.K. Particle deposition indoors: A review. *Indoor Air* **2002**, *12*, 211–214.

45. Heudorf, U.; Neitzert, V.; Spark, J. Particulate matter and carbon dioxide in classrooms-The impact of cleaning and ventilation. *Int. J. Hyg. Environ. Health* **2009**, *212*, 45–55.

46. Slezakova, K.; Castro, D.; Pereira, M.C.; Morais, S.; Delerue-Matos, C.; Alvim-Ferraz, M.C. Influence of tobacco smoke on carcinogenic PAH composition in indoor PM$_{10}$ and PM$_{2.5}$. *Atmos. Environ.* **2009**, *43*, 6376–6382.

47. Alshitawi, M.S.; Awbi, H.B. Measurement and prediction of the effect of students' activities on airborne particulate concentration in a classroom. *Int. J. HVAC R Res.* **2011**, *17*, 446–464.

48. Zhang, Q.; Zhu, Y. Characterizing ultrafine particles and other air pollutants at five schools in South Texas. *Indoor Air* **2012**, *22*, 33–42.

49. Roger, J.; Guinot, B.; Cachier, H.; Mallet, M.; Dubovik, O.; Yu, T. Aerosol complexity in megacities: From size-resolved chemical composition to optical properties of the Beijing atmospheric particles. *Geophys. Res. Lett.* **2009**, *36*, L18806.

50. Gray, H.A.; Cass, G.R.; Huntzicker, J.J.; Heyerdahl, E.K.; Rau, J.A. Characteristics of atmospheric organic and elemental carbon particle concentrations in Los Angeles. *Environ. Sci. Technol.* **1986**, *20*, 580–589.

51. Watson, J.G.; Chow, J.C.; Houck, J.E. PM$_{2.5}$ chemical source profiles for vehicle exhaust, vegetative burning, geological material, and coal burning in northwestern Colorado during 1995. *Chemosphere* **2001**, *43*, 1141–1151.

52. Cachier, H.; Bremond, M.P.; Buat-Menard, P. Carbonaceous aerosols from different tropical biomass burning sources. *Nature* **1989**, *340*, 371–373.

53. Castro, L.M.; Pio, C.A.; Harrison, R.M.; Smith, D.J.T. Carbonaceous aerosol in urban and rural european atmospheres: Estimation of secondary organic carbon concentrations. *Atmos. Environ.* **1999**, *33*, 2771–2781.

54. Robinson, A.L.; Donahue, N.M.; Shrivastava, M.K.; Weitkamp, E.A.; Sage, A.M.; Grieshop, A.P.; Lane, T.E.; Pierce, J.R.; Pandis, S.N. Rethinking organic aerosols: Semivolatile emissions and photochemical aging. *Science* **2007**, *315*, 1259–1262.

55. Dockery, D.W.; Spengler, J.D. Indoor-outdoor relationships of respirable sulfates and particles. *Atmos. Environ.* **1981**, *15*, 335–343.

56. Guo, H.; Morawska, L.; He, C.R.; Zhang, Y.L.; Ayoko, G.; Cao, M. Characterization of particle number concentrations and $PM_{2.5}$ in a school: Influence of outdoor air pollution on indoor air. *Environ. Sci. Pollut. Res. Int.* **2010**, *17*, 1268–1278.

57. Seinfeld, J.H.; Pandis, S.N. *From Air Pollution to Climate Change*; John Wiley & Sons: New York, NY, USA, 2006.

58. Remer, L.A.; Chin, M.; DeCola, P.; Fein-gold, G.; Halthore, R.; Kahn, R.A.; Quinn, P.K.; Rind, D.; Schwartz, S.E.; Streets, D.G.; *et al.* Atmospheric Aerosol Properties and Climate Impacts: Aerosols and Their Climate Effects, 1–2. Available online: http://download.globalchange.gov/sap/sap2-3/sap2-3-final-report-FrontMatter.pdf (accessed on 9 March 2015).

59. Shrivastava, M.K.; Lane, T.E.; Donahue, N.M.; Pandis, S.N.; Robinson, A.L. Effects of gas particle partitioning and aging of primary emissions on urban and regional organic aerosol concentrations. *J. Geophys. Res.* **2008**, *113*, D18301.

60. Huang, Y.; Lee, S.C.; Ho, K.F.; Ho, S.S.H.; Cao, N.Y.; Cheng, Y.; Gao, Y. Effect of ammonia on ozone-initiated formation of indoor secondary product with emissions from cleaning products. *Atmos. Environ.* **2012**, *59*, 224–231.

61. Colome, S.; Kado, N.; Jaques, P.; Kleinman, M. Indoor-outdoor air pollution relations: Particulate matter less than 10 mm in aerodynamic diameter (PM_{10}) in homes of asthmatics. *Atmos. Environ.* **1992**, *26A*, 2173–2178.

62. Ho, K.F.; Cao, J.J.; Harrison, R.M.; Lee, S.C.; Bau, K.K. Indoor/outdoor relationships of organic carbon (OC) and elemental carbon (EC) in $PM_{2.5}$ in roadside environment of Hong Kong. *Atmos. Environ.* **2004**, *38*, 6327–6335.

63. Zhu, C.S.; Cao, J.J.; Shen, Z.X.; Liu, S.X.; Zhang, T.; Zhao, Z.Z.; Xu, H.M.; Zhang, E.K. Indoor and outdoor chemical components of $PM_{2.5}$ in the rural areas of Northwestern China. *Aerosol Air Qual. Res.* **2012**, *12*, 1157–1165.

64. Jones, N.C.; Thornton, C.A.; Mark, D.; Harrison, R.M. Indoor/outdoor relationships of particulate matter in domestic homes with roadside, urban and rural locations. *Atmos. Environ.* **2000**, *34*, 2603–2612.

65. Huang, H.; Zou, C.W.; Cao, J.J.; Tsang, P.K.; Zhu, F.X.; Yu, C.L.; Xue, S.J. Water-soluble ions in $PM_{2.5}$ on the Qianhu campus of Nanchang university, Nanchang city: Indoor-outdoor distribution and source implications. *Aerosol Air Qual. Res.* **2012**, *12*, 435–443.

66. Huang, H.; Zou, C.W.; Cao, J.J.; Tsang, P.K. Carbonaceous aerosol characteristics in outdoor and indoor environments of Nanchang, China, during summer 2009. *J. Air Waste Manag. Assoc.* **2011**, *61*, 1262–1272.

67. Watson, J.G.; Chow, J.C.; Lowenthal, D.H.; Pritchett, L.C.; Frazier, C.A.; Neuroth, G.R.; Robbins, R. Differences in the carbon composition of source profiles for diesel- and gasoline-powered vehicles. *Atmos. Environ.* **1994**, *28*, 2493–2505.

68. Chow, J.C.; Watson, J.G.; Chen, L.W.A.; Arnott, W.P.; Moosmuller, H.; Fung, K.K. Equivalence of elemental carbon by thermal/optical reflectance and transmittance with different temperature protocols. *Environ. Sci. Technol.* **2004**, *38*, 4414–4422.

69. Niu, Z.C.; Wang, S.; Chen, J.S.; Zhang, F.W.; Chen, X.Q.; He, C.; Lin, L.F.; Yin, L.Q.; Xu, L.L. Source contributions to carbonaceous species in PM2.5 and their uncertainty analysis at typical urban, peri-urban and background sites in southeast China. *Environ. Pollut.* **2013**, *181*, 107–114.

70. Kim, E.; Hopke, P.K. Improving source identification of fine particles in a rural northeastern US area utilizing temperature-resolved carbon fractions. *J. Geophys. Res.* **2004**, *109*, 1–13.

71. Cao, J.J.; Lee, S.C.; Ho, K.F.; Fung, K.; Chow, J.C.; Watson, J.G. Characterization of roadside fine particulate carbon and its eight fractions in Hong Kong. *Aerosol Air Qual. Res.* **2006**, *6*, 106–122.

6

A Comparison of ETKF and Downscaling in a Regional Ensemble Prediction System

Hanbin Zhang [1], Jing Chen [2,*], Xiefei Zhi [3] and Yanan Wang [4]

[1] College of Atmospheric Science, Nanjing University of Information & Science Technology, Nanjing 210044, China; E-Mail: zhb828828@163.com

[2] Center of Numerical Weather Prediction of CMA, Beijing 100081, China

[3] Key Laboratory of Meteorological Disaster, Ministry of Education College of Atmospheric Science, Nanjing University of Information & Science Technology, Nanjing 210044, China; E-Mail: xf_zhi@163.com

[4] Center of Meteorological Service of Zhejiang, Hangzhou 310017, China; E-Mail: wangyanan19871120@163.com

* Author to whom correspondence should be addressed; E-Mail: chenj@cma.gov.cn

Academic Editor: Anthony R. Lupo

Abstract: Based on the operational regional ensemble prediction system (REPS) in China Meteorological Administration (CMA), this paper carried out comparison of two initial condition perturbation methods: an ensemble transform Kalman filter (ETKF) and a dynamical downscaling of global ensemble perturbations. One month consecutive tests are implemented to evaluate the performance of both methods in the operational REPS environment. The perturbation characteristics are analyzed and ensemble forecast verifications are conducted; furthermore, a TC case is investigated. The main conclusions are as follows: the ETKF perturbations contain more power at small scales while the ones derived from downscaling contain more power at large scales, and the relative difference of the two types of perturbations on scales become smaller with forecast lead time. The growth of downscaling perturbations is more remarkable, and the downscaling perturbations have larger magnitude than ETKF perturbations at all forecast lead times. However, the ETKF perturbation variance can represent the forecast error variance better than downscaling. Ensemble forecast verification shows slightly higher skill of downscaling ensemble over ETKF ensemble. A TC case study indicates that the overall performance of the two systems are quite similar despite the slightly smaller error of DOWN ensemble than ETKF ensemble at long range forecast lead times.

Keywords: regional ensemble prediction system; initial condition perturbation keyword; ensemble transform Kalman filter; dynamical downscaling

---

1. Introduction

It has long been known that numerical weather prediction (NWP) is sensitive to the initial condition (IC) error, model error and the chaotic nature of atmosphere, thus an ensemble prediction method [1] has emerged as a practical way for providing probabilistic forecasts. Since the ensemble predictions implemented operationally in the early 1990s at the National Centers for Environmental Prediction [2] and at the European Center for Medium-Range Weather Forecast [3], an ensemble prediction system (EPS) has been operational in many meteorological centers to provide operational global weather forecast [4–6].

Since the probability distribution for the various sources of errors are more complicated for regional NWP, it is difficult to predict the meso-scale severe weather. It seems that developing a regional ensemble prediction system (REPS) is a practical way to solve this problem. How to generate IC perturbations for regional ensemble prediction is an important issue. One of the possible choices is dynamical downscaling of a global EPS, which interpolates forecast fields from a set of representative members of the global EPS to obtain different ICs for the regional domain with higher resolution. This method has been successfully applied in some current operational REPS [7–9]. Recently, downscaled IC perturbations have also been successfully applied in experimental convection-permitting REPSs with higher resolution [10,11]. In these convection-permitting ensembles, an intermediate resolution REPS is typically used to transfer the information in a chain of forecasts from the coarse-resolution global EPS to the high-resolution REPS. Although dynamical downscaling is attractive for its simplicity and good performance, the regional small-scale uncertainties cannot be explicitly represented with dynamical downscaling but just following the governance of the global ensemble that is driving it [12]. Some other studies try to generate IC perturbations for REPS by using a regional version of traditional IC perturbation methods, such as the Breeding Growing Mode (BGM), Singular Vectors (SVs), Ensemble Transform Kalman Filter (ETKF), *etc.* It is proved that these methods can also trigger limited ensemble spread and benefit forecast skill for REPS [13–16].

However, so far it is still unclear whether these regional versions of IC perturbation methods, as primarily designed for medium-range forecasting, are fully superior to downscaling when applied to REPS. Bowler and Mylne [17] tested ETKF and downscaling as the IC perturbations generators for a regional version of MetOffice Global and Regional Ensemble Prediction System (MOGREPS), and revealed that the perturbations generated by regional ETKF contain more detail at small scales and less power at large scales with less than 18 h forecast lead time. These perturbations are overall smaller than the ones derived from the downscaling, and the skill of the two ensembles is very similar, with slightly higher skill being seen from the downscaling. Whereas the comparison results of downscaling and regional IC perturbation generators presented by Saito *et al.* [18] are mixed, as the downscaling method tends to perturb synoptic-scale disturbances and showed the best ratio of the ensemble spread to the RMSE, while the regional version of BGM, ET and SVs tend to perturb meso-scale disturbances, thus affecting local intense rains more. Although whether these regional IC perturbation generators can yield advantage over dynamical downscaling is still obscure, there is no doubt that these methods can produce more information of small/meso-scale uncertainties than dynamical downscaling, and this information is particularly useful for the forecasting of local severe convective weather [19,20].

The CMA has been operationally running a REPS since May, 2014, with the IC perturbations generated by a regional version of ETKF. Since this operational REPS is coupled with an operational global ensemble prediction system (GEPS) of CMA, it is also possible for the REPS to obtain IC perturbation from dynamical downscaling of global ensemble perturbations. This paper will give a detailed comparison of ETKF and downscaling in the operational REPS environment. In the study we hope to have a further understanding of the advantages and disadvantages of the two typical kinds of IC perturbation methods, and this investigation is also expected to provide some information for the improvement of the present REPS in the future.

The outline of this article is as follows. Section 2 describes the ensemble forecasting system, the structure of the IC perturbations, and the investigation set-up. Section 3 presents the results and discussion from the evaluation of perturbation quality and various ensemble forecast quality measures. Afterwards, the Typhoon track forecast quality is investigated. A summary and conclusions of the obtained results is provided in Section 4.

2. System and Method

2.1. Introduction of the Regional Ensemble Prediction System

The REPS in CMA has been running operationally so far, this system with relatively high-resolution aims at providing probabilistic forecast for meso/small-scale severe weather phenomenon, such as heavy precipitation and tropical cyclone (TC).

This REPS is constructed based on a regional model of GRAPES-Meso (regional version of Global and Regional Assimilation and Prediction System) [21]. The GRAPES-Meso regional model runs on a regular latitude-longitude grid with a resolution of 0.15 degree in the horizontal and 33 model levels vertically. Model domain is set to 70–145.15°E, 15–64.35°N, covering the whole area of China. The IC perturbations of this REPS (abbreviated as GRAPES-REPS) are generated by a regional ETKF approach, and the model uncertainty is represented by multiple physics [4]. Since the GRAPES-REPS have coupled with the GEPS of CMA, the lateral boundary conditions (LBCs) of the GRAPES-REPS are also perturbed. Figure 1 shows the model domain of GRAPES-REPS that nested in the global ensemble. The GRAPES-REPS consist of 15 members, including a control run and 14 perturbed ensemble members. In each day the system started at 0600 UTC, 1200 UTC, 1800 UTC and 0000 UTC. For each start time the system provide 6 h forecast perturbations for the next ETKF cycle, specifically for 1200 UTC and 0000 UTC initiate time the model integrate to 72 h to provide ensemble forecast products. Five variables (zonal wind u, meridional wind v, potential temperature θ, Exner pressure π and specific humidity q) in the ICs are perturbed.

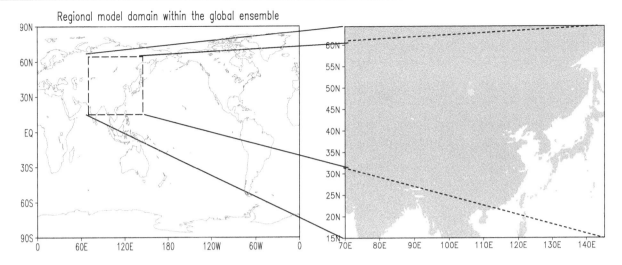

Figure 1. Model domain of GRAPES-REPS within a global ensemble.

2.2. Introduction of the IC Perturbation Schemes

In our study, two IC perturbation schemes are used, one is regional ETKF (as applied in the operational run), and the other is dynamical downscaling of the GEPS.

2.2.1. ETKF

Wang and Bishop [22] demonstrated the viability and effectiveness of ETKF in generating IC perturbations (as also called analyses perturbations) for global ensemble forecast. The ETKF can generate IC perturbations that correspond well to the observational density. In addition, compared with BGM technique that maintain variance in few directions, ETKF can maintain comparable amounts of variance in all orthogonal and uncorrelated directions spanning its ensemble perturbation subspace; moreover, the computation cost of ETKF is not much more than BGM.

The derivation of ETKF analyses perturbations is based on the hypothesis that the forecast covariance matrices and analysis covariance matrices can be represented by forecast perturbations X^f and analysis perturbations X^a. The relationship between X^f and X^a is established after solving the optimal data assimilation equation; as a result, the forecast perturbation can be transformed to analysis perturbation through a transformation matrix T, that is

$$X^a = X^f T \Pi \tag{1}$$

where forecast perturbations are listed as columns in the matrix X^f and analysis perturbations are listed as columns in the matrix X^a. Π is a scalar inflation factor to inflate the analysis perturbation amplitude so as to ensure that the 6 h forecast ensemble variance is consistent with the control forecast error variance. Following Wang and Bishop, T is given by

$$T = C(\Gamma + I)^{-1/2} C^T \tag{2}$$

where columns of the matrix C and Γ contain the eigenvectors and the corresponding eigenvalues of the matrix:

$$(Z^f)^T H^T R^{-1} H Z^f \tag{3}$$

where the matrix H is the linear observation operator that maps model variables to observed variables and the matrix R is the observation error covariance matrix.

2.2.2. Dynamical Downscaling

A traditional "downscaling" mechanism is a process that aims at finding a mathematical relation between the global and local fields, that is stable in time and, valid for a variety of different meteorological systems. The generation of some systems (e.g., convective phenomenon) are too complicated to be linearly predictable from the large-scale flow, so the downscaling usually achieved by a complicated, nonlinear "transfer function" [23]. There are a variety of downscaling techniques in the literature, but two major approaches can be identified at the moment, namely, dynamical downscaling and empirical (statistical) downscaling. Dynamical downscaling approach is a method of extracting local-scale information by regional models with the coarse global data used as boundary conditions [24]. For a regional ensemble forecast, the dynamical downscaling of an ensemble of global IC stats will produce the IC stats of regional ensemble [25]. Generally, this downscaling approach to generate IC perturbations for REPS is attractive due to its relative simplicity and practicality of implementation.

The resolution of the GEPS in CMA is T639L60 (spectral triangular T639 with 60 vertical levels, corresponding to 30 km resolution). A masked breeding method [26] is applied to compute ICs of this GEPS, with 12 h optimization time interval. For each initial time, the forecast perturbations of the pervious breeding cycle are scaled to have initial amplitude comparable to an estimate of the analysis error. Mathematically,

$$X_{ij}^a = X_{ij}^f c_{ij} \tag{4}$$

where X^a_{ij} and X^f_{ij} are analysis perturbations and forecast perturbations at latitude i and longitude j, c_{ij} is a rescaling factor which is a function of latitude and longitude.

The background state as well as the LBCs of GRAPES-REPS is provided by a GEPS. This configuration enable IC perturbations of GRAPES-REPS be obtained from dynamical downscaling of this GEPS. This is achieved by interpolating the IC stats of GEPS to the 0.15 by 0.15 degree resolution through the initialization process of GRAPES-Meso regional model.

2.3. Experimental Set-Up

In the present study, the ETKF method and the downscaling method are compared using the same unperturbed analysis and the same forecast model. The numerical experiments are based on the operational GRAPES-REPS of CMA. The regional ensemble was run twice to produce forecasts, one set of forecasts generated IC perturbations by the regional ETKF, and the other was run using a downscaling of the global perturbations as IC perturbations.

The two different ensembles compared in this work will be denoted as ETKF and DOWN. The one-month period of 1 August 2012 to 31 August 2012 is chosen to conduct this comparison test, with both sets of ensembles initiated at 1200 UTC each day. Forecasts were evaluated to a lead-time of 72 h. The system settings of the two tests are identical to the operational run.

The background state and the LBCs of experimental REPS are provided by T639 global ensemble forecast data. The T639 global analysis states corresponding to each forecast lead times of the forecasts are interpolated to a common regular 0.15 by 0.15 degree resolution to verify the forecasts of upper air weather variables. The observational TC track data are provided by the Joint Typhoon Warning Center (JTWC) Best Track.

3. Results and Discussion

We now evaluate the quality of IC perturbation states and ensemble forecasts from each methodology. For a regional ensemble forecast, it is desirable to provide information of all scales, not only synoptic scales but also convective scales, therefore we start with an examination of the scale characteristics of ETKF perturbations and DOWN perturbations. Next, since the ensemble spread growth is closely correlated with the perturbation growth, we investigate how the two types of perturbations evolve, and how the perturbation amplitude increase with lead time. In addition, an "ensemble perturbation precision test" is conducted to determine whether the two perturbation techniques can better represent the forecast errors (e.g., locations where perturbation amplitude is large corresponds to locations where forecast error is large). Thereafter, we statistically evaluate the ensemble forecast skill, with a series of probability verification scores used. We will also present a study to evaluate the practicability of two methods in a particular weather case.

3.1. Power Spectra Analysis

A good ensemble forecast can provide sufficient uncertainty information, either for small-scale phenomenon or for large-scale phenomenon. The spatial scale characteristics of two types of perturbations are investigated, this is achieved by calculating the power spectra of both the ETKF and DOWN perturbations. A 2-dimemsional Discrete Cosine Transform (2D-DCT) [27], which is suitable for spectral analysis of data on a limited area, is used to conduct power spectra analyses. We first find the difference between each perturbed forecast and the ensemble mean, and then calculate the power spectra by 2D-DCT. The power spectra are calculated for each ensemble member, to create an average power spectra value.

Figure 2 shows power spectra of 500 hPa temperature perturbations as a function of wavelength, for both ETKF perturbations and DOWN perturbations. The power spectra of initial perturbations and 6 h forecast perturbations are presented. Results from initial (00 h) forecasts (Figure 2a) show that the power of the ETKF perturbations is greater than that of the downscaling perturbations at wavelengths less than 1100 km. In particular, for wavelengths less than 60 km (around two grid lengths of the T639 global model), there is no power for the global ensemble perturbations. These scales cannot be better resolved by the global model, so the perturbations derived from the global ensemble exhibit less power at these length-scales. Whereas for scales over 1100 km more power can be found in DOWN ensemble, as the maximum power can reach to 40 k^2 (corresponding to wavelength of 5000 km), while the maximum value for ETKF ensemble can only reach to 20 k^2 (corresponding to wavelength of 3200 km). The results from 12 h forecasts (Figure 2b) show that the scale characteristics of the two perturbation schemes get closer with forecast lead time. The 12 h ETKF perturbations exhibit amplified power than that of 00 h at larger scales, and the most powerful scale is 5000 km, with the power value of

60 K^2; while the 12 h downscaling perturbations have increased power than that of 00 h at all length-scales, including the small-scales that cannot be resolved by downscaling perturbations at initial time.

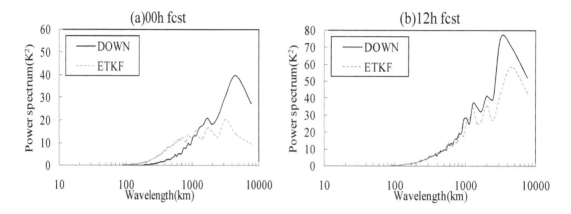

Figure 2. All member averaged power spectra of 500 hPa temperature perturbations as a function of wavelength for ETKF and DOWN. (**a**) Initial time; (**b**) 12 h forecast lead time.

The results presented above indicate, when applied to the self cycling of GRAPES-REPS, the ETKF technique can create analysis perturbations from forecast perturbations which are completely produced by regional model and hence, in principle, provide IC perturbations at all scales resolved by regional model. The greater power of small-scales can enable the ETKF perturbations to better capture the convective, high impact weather uncertainty, whereas the downscaling perturbations could not represent the small-scale uncertainty at the initial time, and are better than ETKF at representing the large-scale uncertainty.

3.2. Perturbation Growth Characteristics

It has just been shown that the ETKF perturbations have greater power at smaller scales while the DOWN perturbations have greater power at larger scales. In order to investigate the perturbation characteristic intuitively, an attempt has been made to account for the distribution and evolution characteristics of both kinds of perturbations. This is achieved by calculating an "approximate energy norm" [28], defined as

$$\frac{1}{2}\left[u'^2(i,j,k)+v'^2(i,j,k)\right]+\frac{c_p}{T_r}T'^2(i,j,k) \tag{5}$$

where u', v' and T' are wind and temperature perturbations, c_p is the specific heat and T_r is the reference temperature.

Figure 3 shows horizontal distributions of the energy norm averaged over all members at all levels, for both ETKF (left panel) and downscaling (right panel) ensembles, respectively. For the ETKF ensemble, the energy norm of analysis perturbations (Figure 3a) in eastern China is generally lower than that in the plateau and Western Pacific regions, due to the larger number of observations in this region. For the downscaling ensemble, the energy norm of analysis perturbations (Figure 3b) distribution do not show obvious observation impact because the rescaling factors in the global masked breeding are designed empirically from climatology data, although the global observation distribution in the mask is considered, the breed perturbations cannot reflect the regional observations distribution in detail. Additionally, the ETKF analysis perturbations exhibit more small-scale characteristic than DOWN analysis perturbations, while the DOWN perturbation pattern is larger in scale, and this difference can also reflect the analysis result in Section 3.1.

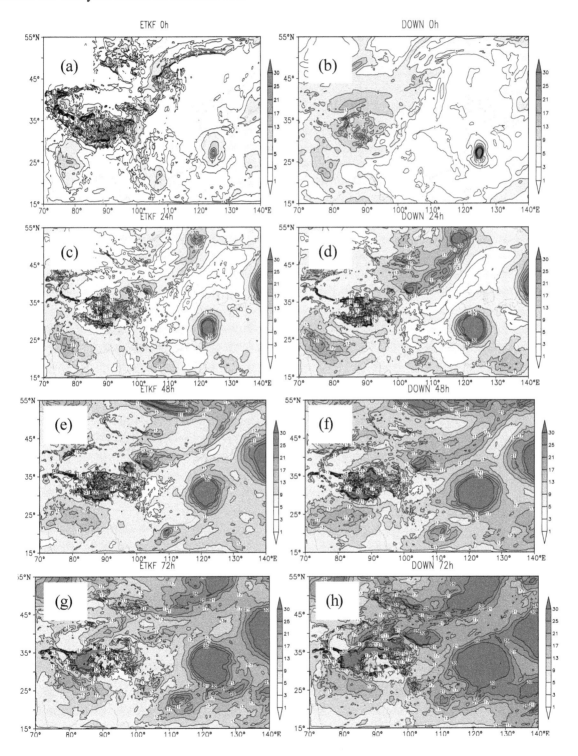

Figure 3. All members vertically averaged perturbation energy norm (unit:J/kg) at different forecast lead times, for ETKF and DOWN, respectively. (**a**) ETKF 00 h; (**b**) DOWN 00 h; (**c**) ETKF 24 h; (**d**) DOWN 24 h; (**e**) ETKF 48 h; (**f**) DOWN 48 h; (**g**) ETKF 72 h; and (**h**) DOWN 72 h.

For 24 h forecast lead time (Figure 3c,d), the perturbations for both ensembles show remarkable growth, especially for DOWN perturbations in Western Pacific regions. With the increase of forecast lead time, the perturbation patterns for both ensembles become similar. Take 48 h (Figure 3e,f) and 72 h (Figure 3g,h) forecast lead times for example, the large perturbations regions in both ETKF and DOWN perturbation states correspond very well. The results indicate that the difference between the two systems in the perturbation distribution pattern for short range forecast is more significant than that of long range forecast.

Aside from the perturbation distribution pattern, it is desirable to compare both ensembles in terms of perturbation magnitude. We average the perturbation energy norm at all grid points at each level to get the vertical distributions of energy norm. Figure 4 illustrates such energy norm profiles for 0–36 h forecast lead times. The energy norm of ETKF ensemble (Figure 4a) can keep steady growth with forecast lead time, and the most remarkable growth level is around 250 hPa. For example, at 6 h lead time, the energy norm at 250 hPa is 2.7 J/kg, while at 36 h lead time, the corresponding value is 4 J/kg. The energy value at higher or lower levels is relatively smaller. For DOWN ensemble, the energy norm growth characteristics are similar. The largest energy norm can also be found at 250 hPa, with a value of 3.4 J/kg for 6 h forecast lead time and it 4.4 J/kg for 36 h lead time. It should be noted that the energy norm of DOWN ensemble perturbations is always larger than that of the ETKF ensemble perturbations at corresponding levels and corresponding lead times.

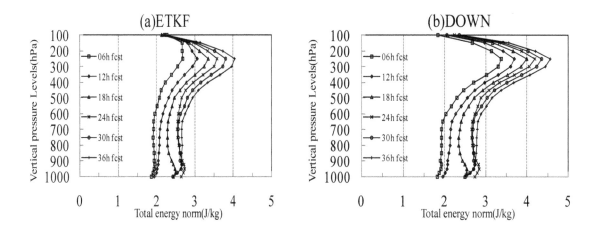

Figure 4. Vertical distributions of ensemble mean total energy (unit: J/kg), different lines denote different forecast lead times. (**a**) ETKF; (**b**) DOWN.

3.3. Ensemble Perturbation Precision Test

While larger perturbation values can indicate the larger magnitude of spread (as larger spread is always desirable for ensemble forecast), it is also interesting and important to investigate the "ensemble perturbation precision". Studies by Toth *et al.* [29] and Zhu *et al.* [30] suggest that the ability of an ensemble to predict case-dependent forecast uncertainty is a critical criterion to evaluate an ensemble forecast system. The true forecast error variance can be regarded as a random variable around the ensemble perturbation variance. An accurate prediction of forecast error variance is one in which the true forecast error variance distributes closely to the ensemble perturbation variance; that is, the difference of the forecast error variance around the ensemble perturbation variance is small. We refer to the ability of an ensemble to get forecast error variance right on every day at every grid point variable as "the precision" of the ensemble perturbation variance [22]. Information about the degree of ensemble variance precision can be used to increase the accuracy of error probability density functions derived from ensemble variances. Here, we introduce tests of ensemble variance precision.

To analyze how well the ensemble perturbation variance can explain the forecast error variance, we follow the method used by Wang and Bishop [22] and Wei *et al.* [6]. First, we compute the ensemble perturbation variance and squared error of a variable at each grid point of a particular pressure level. A scatterplot (which is not shown) can then be drawn by using ensemble perturbation variance (abscissa) and forecast errors variance for all grid points. Since the grid points for the both REPSs are 502×330, we next divide the points into 330 equally populated bins (with 502 grid points in each bin) in order of increasing ensemble variance. The ensemble and forecast variances are then averaged within each bin. It is the averaged values from each bin that are plotted. The relationship between ensemble perturbation variance and forecast error variance of 500 hPa temperature for ETKF and DOWN ensembles in forms of such plot are shown in Figure 5. If the number of bins is reduced (e.g., 33 bins with 5020 grid points in each bin), it is expected that the curve will be smoother. The result from 33 bins is shown by a solid line. For 6 h forecast (Figure 5a,b), the results from the 330-bin case (dotted line) show that the range of forecast error variance (maximum minus minimum values) explained by the ensemble variance is larger for ETKF(3.3 K^2) than DOWN (0.4 K^2). For 33-bin case (solid line), the range of forecast error variance explained by ETKF ensemble variance is also larger (1.5 K^2) than that of DOWN (0.3 K^2). This shows that ETKF perturbations are better than DOWN perturbations at being able to distinguish times and locations where forecast errors are likely to be large from the times and locations where forecast errors are likely to be small. For 60 h lead time (Figure 5c,d), the range of forecast error variance (10.8 K^2 from 330-bin case and 8.3 K^2 from 33-bin case) explained by ETKF ensemble variance is still larger than that of DOWN ensemble (9.5 K^2 from 330-bin case and 7.5 K^2 from 33-bin case), but the difference between the two ensembles become smaller, compared to the 6 h forecast.

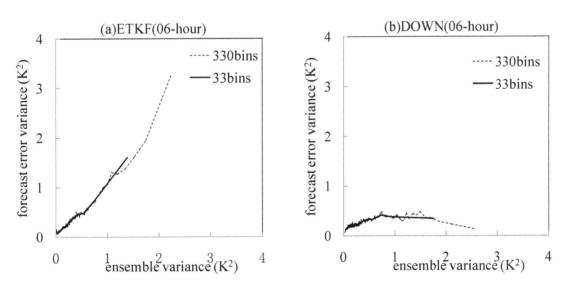

Figure 5. *Cont.*

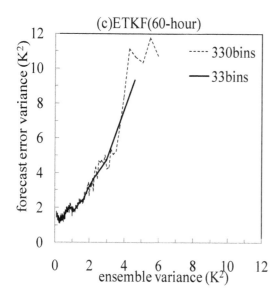
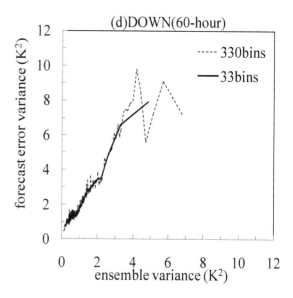

Figure 5. Relationship between the 500 hPa temperature ensemble perturbation variance and forecast error variances with the value averaged from each of 33 bins (solid lines) and 330 bins (dotted lines) at all grid points. (**a**) ETKF 6 h forecast; (**b**) DOWN 6 h forecast; (**c**) ETKF 60 h forecast; (**d**) DOWN 60 h forecast.

3.4. Ensemble Verification Results

To compare the results of both perturbation methods, we have used several verification methods. The methods are root mean square error (RMSE) for ensemble mean, the ensemble spread, the continuous ranked probability skill score (CRPS), and the Talagrand diagram. The results are reported for several model output variables.

3.4.1. Root Mean Square Error and Ensemble Spread

A useful measure of the skill of an ensemble prediction system is how well the variation in the spread (deviation of the ensemble about its mean) of the ensemble matches the variation in the RMSE of the ensemble mean forecast. We calculated the spatially averaged spread of all variables at all levels for each lead time, and compared this to a spatially averaged RMES of the ensemble mean. Figure 6 shows one-month averaged ensemble mean RMSE and ensemble spread for ETKF ensemble and DOWN ensemble, with two upper air variables of 500 hPa wind speed (WS500), 850 hPa temperature (T850) and one near surface variable of 10 m U wind (U10m) presented for comparison. It turns out that for all these variables, both methods have a lack of spread, and spread growth is slower than RMSE growth for both ensembles. Overall, the DOWN ensemble shows larger spread at all forecast lead times comparatively, with relatively smaller RMSE. Similar results can also be observed for other variables at different pressure levels (not shown). The results suggest that the larger spread of DOWN ensemble can really enhance the accuracy of ensemble mean forecast.

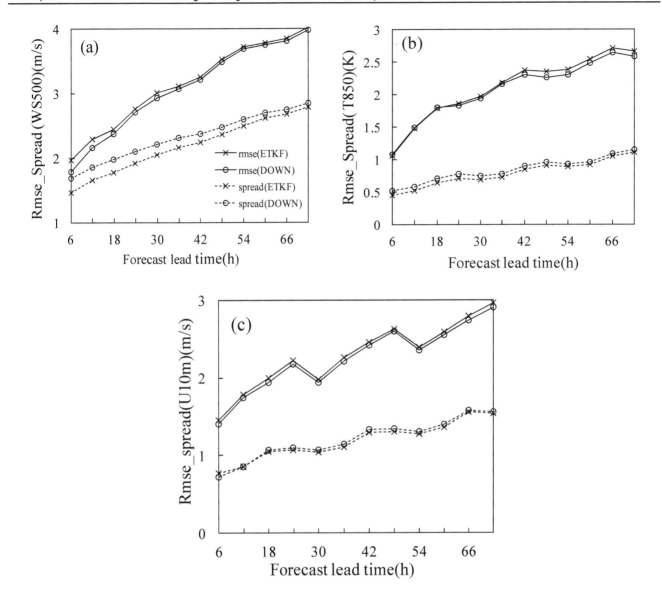

Figure 6. RMSE of ensemble mean and ensemble spread for ETKF and DOWN, respectively. (**a**) 500 hPa wind speed; (**b**) 850 hPa temperature; (**c**) 10 m U wind.

3.4.2. Continuous Rank Probability Score

The continuous rank probability score (CRPS) is an overall measure of the skill of a probabilistic prediction, measuring the skill of the ensemble mean forecast as well as the ability of the perturbations to capture the deviations around that. For a variable x, if the ensemble predicted probability density function is $p(x)$ and the observational value is x_o, the CRPS is given by [31]

$$CRPS = \int_{-\infty}^{\infty} \left[P(x) - H(x - x_o) \right]^2 dx \tag{6}$$

where

$$P(x) = \int_{-\infty}^{x} p(y) dy \tag{7}$$

and $H(x)$ is the step function with:

$$H(x) = \begin{cases} 0(x < 0) \\ 1(x >= 0) \end{cases} \qquad (8)$$

The CRPS is a penalty score, so smaller values are better. Figure 7 shows the CRPS of WS500, T850 and WS850 for both ETKF ensemble and DOWN ensemble. It is clear that the DOWN ensemble performs better (with smaller CRPS value) than ETKF for all the variables within 72 h lead times. CRPS verification on other variables can give similar results (not shown). This demonstrates that the DOWN method is better at providing probabilistic forecast than ETKF. Note, however, that the advantage of DOWN ensemble is quite limited, and the overall performances of the two ensembles are quite similar.

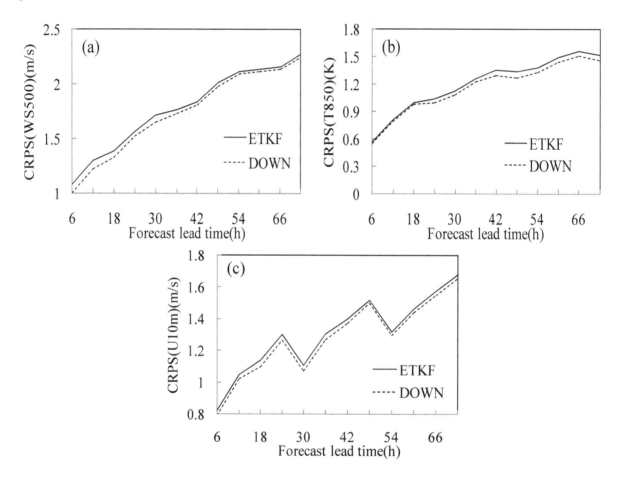

Figure 7. CRPS as a function of forecast lead time, for ETKF and DOWN, respectively. (**a**) 500 hPa wind speed; (**b**) 850 hPa temperature; (**c**)10 m U wind.

3.4.3. Talagrand Diagram

Another measure of statistical reliability is the Talagrand diagram [32]. This is the statistic of the frequency that the observation lays inside or outside the whole ensemble. A more reliable EPS should have a more flat pattern. "U" shape means lack of spread, "J" or "L" shapes mean there is bias in the system. Figure 8 shows the Talagrand diagram for WS500, T850 and U10m for 24 h lead time. It is evident that for all the graphs, the diagram of DOWN ensemble is closer to flat than the ETKF ensemble; this indicate that the frequency that observations lay inside the whole ensemble is higher for DOWN than ETKF. From Figure 6 we know that both ensembles are under-spread, thus, comparatively speaking, this more flat pattern of DOWN ensemble is desirable.

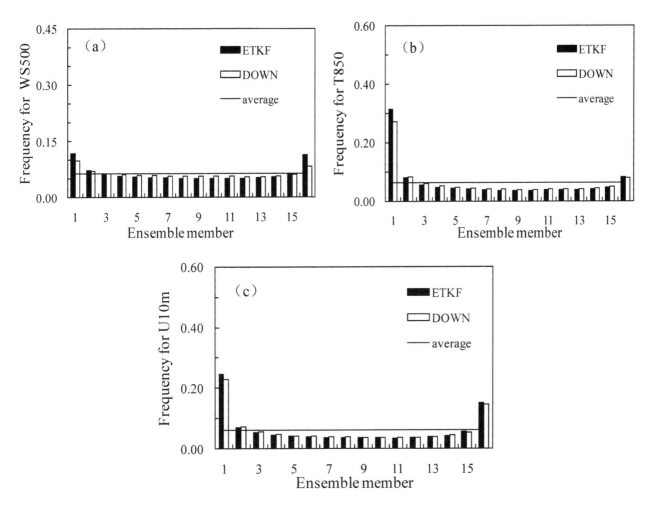

Figure 8. Talagrand for ETKF and DOWN at forecast lead time of 24 h. (**a**) 500 hPa wind speed; (**b**) 850 hPa temperature; (**c**)10 m U wind.

3.5. A TC Case Study

A TC case is studied to assess the practical performances of both ensembles. The 11th typhoon "Haikui" (1211) in 2012 was one of the most severe threats to Mainland China and caused several hazardous disasters, including heavy precipitation, windstorms, and storm surges. Haikui was born in the Central Pacific (about 140.7°E, 23.2°N) at 08030000 UTC and landed on Xiangshan in Zhejiang Province, China at 08071920 UTC (Figure 9). Graded as "severe typhoon" according to the CMA, Haikui was characterized by generation in the high latitude area, rapid intensification just before landing, and stagnation after landing. Here the track forecasts of Haikui (1211) within 72 h lead time from ETKF ensemble and DOWN ensemble are compared.

The predicted tracks of Haikui from ensemble members and ensemble mean of ETKF ensemble and DOWN ensemble are displayed in Figure 10. All forecasts are initiated at 08061200 UTC with a forecast length of 72 h. From the ETKF ensemble forecast (Figure 10a), we note that the tracks forecasted by all the members exhibit a divergence characteristic; this can reflect the spread growth with forecast lead time. The forecasted tracks from several ensemble members are very close to the observation, except the ones with drastic northward turning. From DOWN ensemble forecast (Figure 10b), it is clear that there are also some members that correspond well to the observation, and it seems that the ensemble mean forecast of DOWN is better than that of ETKF. A significant contrast of ETKF ensemble and DOWN ensemble is the variability of the directions between the forecasted tracks, since all the tracks forecasted by ETKF ensemble can pass through the rectangular region, while the tracks forecasted by DOWN ensemble are more diverged with some tracks laying outside of the rectangular region. Although this more dispersive distribution of DOWN forecasted tracks may possible to comprise the true TC course, such great spread of tracks might not be considered desirable as it can also brought confusing for users.

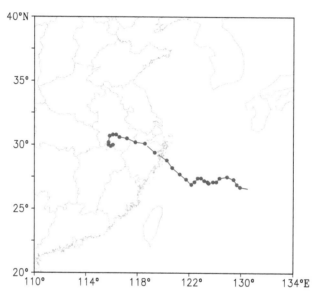

Figure 9. Observed Track of Haikui (1211).

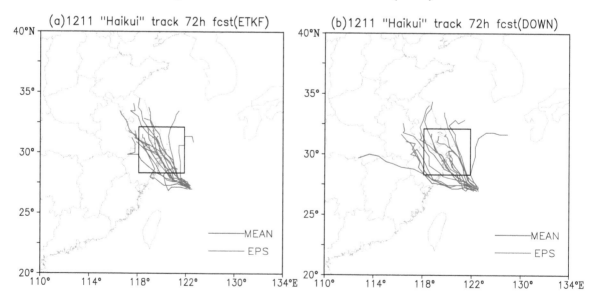

Figure 10. Seventy-two hour track forecasts of Haikui (1211) from ensemble forecasts of (**a**) ETKF and (**b**) DOWN. Red line is ensemble mean forecast (MEAN) and blue lines are ensemble member forecasts (EPS).

An objective of ensemble forecast is to improve the forecast accuracy. To assess the TC track forecast accuracy of the two ensembles, the track forecast error is studied. At one forecast lead time, the TC track forecast error is defined to be the distance of the forecasted track location from the observational location. Such forecast error in terms of track distance is displayed in Figure 11. As shown in Figure 11a, the 36 h forecast error of all ensemble members are limited within a range of 300 km, and the error of ensemble mean is 73 km at 36 h lead time, which is significantly less than most member forecasts. However, we can also find a significant error growth beyond 36 h lead time, and some members exhibit extremely large error, for example, there are three members whose error values are over 600 km at 72 h lead time. The 72 h forecast error of ensemble mean can still maintain a reasonable value of 387 km. As shown in Figure 11b, the error characteristic of DOWN ensemble is very similar to that of ETKF ensemble within 36 h forecast lead time, and the error of ensemble mean is small (66 km for 36 h forecast). After 36 h forecast, it is obvious that the error value of DOWN ensemble member forecasts cover a wider range than ETKF ensemble. The 72 h forecast error value of some members can reach 1160 km, while some members can exhibit very small errors (smaller than 50 km). The error value of ensemble mean is 363 km, which is a little less than that of ETKF.

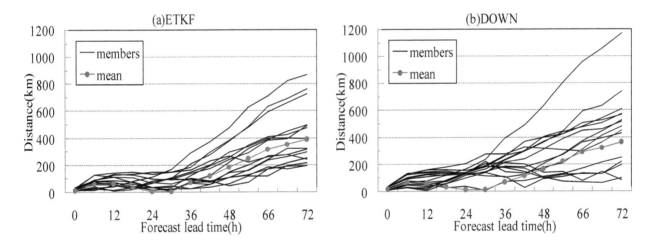

Figure 11. Distance of forecasted tracks and the observational track of Haikui (1211) as a function of forecast lead time, for (**a**) ensemble members and ensemble mean of ETKF and (**b**) ensemble members and ensemble mean of DOWN, respectively.

Generally speaking, the overall skills of the ensemble mean forecast of the two ensembles are very similar, and the DOWN ensemble has slightly higher skill for long forecast lead times. From this TC case study, it has not been possible to see clearly improved performance from either ensemble.

4. Summary and Conclusions

Based on the operational GRAPES-REPS, this paper carried out comparative studies on two IC perturbation schemes for regional ensemble, namely regional ETKF and dynamical downscaling. Using the two IC perturbation schemes, two consecutive ensemble forecast tests in an operational environment are conducted for a period of one month. The perturbation characteristics are investigated, meanwhile ensemble verification are implemented by use of several probability forecast verification methods. Additionally, a TC case of "Haikui" is studied to examine the practical effectiveness of the two regional ensemble forecasts. The main conclusions of this research are as follows:

The perturbations generated by ETKF have larger power at smaller scales, while the perturbations have more power at larger scales. The small-scale cannot be well resolved by the global model, so it is not surprising that the downscaling perturbations have a lack of small-scale power within short forecast range. With the forecast lead time increasing, the differences between ETKF and downscaling perturbations at different scales decrease.

The distribution pattern of both types of perturbations are quite different at initial time, the ETKF perturbations can reflect observation density while the downscaling distribution cannot. The two types of perturbations showed different growth characteristics, this is mainly due to the different ways in which the perturbations are calculated. The perturbations of the global ensemble are generated using a masked BGM, and the global perturbations at 12 h forecast lead time are rescaled by a rescaling factor, so that the perturbations can match approximately the global model forecast error at 12 h forecast lead time in the next cycle. When this is used to drive the IC perturbations for the regional model, the difference between the perturbed forecasts and the control forecast at initial time is relatively larger. When perturbations are calculated using the regional ETKF, the 6 h forecast perturbations are transformed and rescaled, and the perturbation magnitude will match the regional model forecast error at 6 h forecast lead time in the next ETKF cycle. Since there is significant difference between the calculated regional model 6 h forecast error and global model 12 h forecast error, the perturbation magnitude of the ETKF ensemble can be much less than that of the corresponding downscaling ensemble.

Although the downscaling perturbations shows better growth, these perturbations show no more skill than ETKF perturbations on representing forecast error. The range of forecast error variance explained by the ensemble variance is larger for ETKF than downscaling. This result indicate that the ETKF perturbations are better than downscaling perturbations at being able to distinguish times and locations where forecast errors are likely to be large from the times and locations where forecast errors are likely to be small.

The one-month statistics of ensemble verifications are also indicative. The RMSE and spread from the two ensembles show that the DOWN ensemble has slightly higher skill than the ETKF ensemble, and the results from CPRS and Talagrand histogram also support this conclusion. It is known that these scores can be influenced by the magnitude of the initial spread [5]. As Section 3.2 shows, the initial spread of DOWN ensemble is generally larger. This may reduce the forecast error of the DOWN ensemble. A conclusion will be drawn from the future comparison with both systems having similar initial spread.

A TC case study shows that the difference in the ensemble spread between the two systems dominate much of the comparison, the tracks from ETKF ensemble forecast are more concentrated, while the tracks from DOWN ensemble forecast are more dispersive. The track forecast error is also compared to assess whether there is any potential benefit from either system. It seem that the overall skill of the two systems are quite similar despite slightly smaller error of DOWN ensemble than ETKF ensemble at long range forecast lead times.

Now that a comprehensive comparison between IC perturbation methods of regional ETKF and dynamical downscaling has been made, one may revisit the question of whether it was right to implement this downscaling in the operational REPS of CMA, or if there is another way to improve the present ETKF based REPS. The results presented here indicate that the performance of the downscaling perturbations ensemble is better than that of the regional ETKF perturbations ensemble in some respects, such as the larger spread growth and better ensemble verification scores. Moreover, as described in Bowler and Mylne [15], the perturbations derived from downscaling are more consistent with the lateral boundary. However, there is no doubt that the downscaling perturbations have a lack of small-scale information, and the ETKF ensemble perturbations can better represent the forecast error of regional model. Although the above results are mixed, these comparisons may shed some light on further improvement in present system. A practical way to take advantages of both ensemble perturbations is developing a blending technique, which has been primarily investigated by Caron [33] and Wang *et al.* [34]. This technique can obtain blended perturbations that contain large-scale component from downscaling perturbations and small-scale component from ETKF perturbations. As this blending technique can take the advantages of both downscaling perturbations and regional ETKF perturbations, it is expected to be an appropriate way to achieving good ensemble forecast for the CMA REPS.

Acknowledgments

This work is financially supported by grants from the National Natural Science Foundation of China (Grant No. 91437113), the Special Fund for Meteorological Scientific Research in the Public Interest (Grant Nos. GYHY201506007 and GYHY201006015), the National 973 Program of China (Grant Nos. 2012CB417204 and 2012CB955200) and the Scientific Research & Innovation Projects for Academic Degree Students of Ordinary Universities of Jiangsu (Grant No. KYLX_0827).

Author Contributions

Hanbin Zhang conceived and designed the experiment. The paper was written by Hanbin Zhang with a significant contribution by Jing Chen and Xiefei Zhi. Yanan Wang analyzed the data and presented the results.

Conflicts of Interest

The authors declare no conflict of interest.

References

1. Leith, C.E. Theoretical skill of Monte Carlo forecasts. *Mon. Wea. Rev.* **1974**, *102*, 409–418.
2. Toth, Z.; Kalnay, E. Ensemble forecasting at NMC: The generation of perturbations. *Bull. Am. Meteor. Soc.* **1993**, *74*, 2317–2330.
3. Molteni, F.; Buizza, R.; Palmer, T.N.; Petroliagis, T. The ECMWF ensemble prediction system: Methodology and validation. *Quart. J. R. Meteor. Soc.* **1996**, *122*, 73–119.
4. Houtekamer, P.L.; Lefaivrem, L.; Derome, J.; Ritchie, H.; Mitchell, H.L. A system simulation approachto ensemble prediction. *Mon. Wea. Rev.* **1996**, *124*, 1225–1242.
5. Buizza, R.; Houtekamer, P.L.; Toth, Z.; Pellerin, P.; Wei, M.; Zhu, Y. A comparison of the ECMWF, MSC and NCEP global ensemble prediction systems. *Mon. Wea. Rev.* **2005**, *133*, 1076–1097.

6. Wei, M.; Toth, Z.; Wobus, R.; Zhu, Y.; Bishop, C.H.; Wang, X. Ensemble Transform Kalman Filter-based ensemble perturbations in an operational global prediction system at NCEP. *Tellus* **2006**, *58A*, 28–44.

7. Marsigli, C.; Boccanera, F.; Montani, A.; Paccagnella, T. The COSMO-LEPS mesoscale ensemble system: Validation of the methodology and verification. *Nonlinear Processes Geophys.* **2005**, *12*, 527–536.

8. Frogner, I.L.; Haakenstad, H.; Iversen, T. Limited-area ensemble predictions at the Norwegian Meteorological Institute. *Quart. J. R. Meteor. Soc.* **2006**, *132*, 2785–2808.

9. Bowler, N.E.; Arribas, A.; Mylne, K.R.; Robertson, K.B.; Beare, S.E. The MOGREPS short-range ensemble prediction system. *Quart. J. R. Meteor. Soc.* **2008**, *134*, 703–722.

10. Hohenegger, C.; Walser, A.; Langhans, W.; Schär, C. Cloud-resolving ensemble simulations of the August 2005 Alpine flood. *Quart. J. R. Meteor. Soc.* **2008**, *134*, 889–904.

11. Peralta, C.; Ben Bouallegue, Z.; Theis, S.E.; Gebhardt, C.; Buchhold, M. Accounting for initial condition uncertainties in COSMO-DE-EPS. *J. Geophys. Res.* **2012**, *117*, 1–13.

12. Wang, Y.; Bellus, M.; Wittmann, C.; Steinheimer, M.; Weidle, F.; Kann, A.; Ivatek-Sahdan, S.; Tian, W.; Ma, X.; Tascu, S.; *et al.* The Central European limited area ensemble forecasting system: ALADIN-LAEF. *Quart. J. R. Meteor. Soc.* **2011**, *134*, 483–502.

13. Stensrud, D.J.; Brooks, H.E.; Du, J.; Tracton, M.S.; Rogers, E. Using ensembles for short-range forecasting. *Mon. Wea. Rev.* **1999**, *127*, 433–446.

14. Du, J.; DiMego, G.; Tracton, M.S.; Zhou, B. NCEP short range ensemble forecasting (SREF) system: Multi-IC, multi-model and multi-physics approach. *CAS/JSC WGNE Res. Act. Atmos. Ocea. Modell.* **2003**, *33*, 5.09–5.10.

15. Li, X.; Charron, M.; Spacek, L.; Candille, G. A regional ensemble prediction system based on moist targeted singular vectors and stochastic parameter perturbations. *Mon. Wea. Rev.* **2008**, *136*, 443–462.

16. Zhang, H.; Chen, J.; Zhi, X.; Li, Y.; Sun, Y. Study on the application of GRAPES regional ensemble prediction system. *Meteorol. Mon.* **2014**, *40*, 1077–1088.

17. Bowler, N.E.; Mylne, K.R. Ensemble transform Kalman filter perturbations for a regional ensemble prediction system. *Quart. J. R. Meteor. Soc.* **2009**, *135*, 757–766.

18. Saito, K.; Hara, M.; Seko, H.; Kunii, M.; Yamaguchi, M. Comparison of initial perturbation methods for the mesoscale ensemble prediction system of the Meteorological Research Institute for the WWRP Beijing 2008 Olympics Research and Development Project (B08RDP). *Tellus* **2011**, *63*, 445–467.

19. Chen, J.; Xue, J.; Yan, H. A new initial perturbation method of ensemble mesoscale heavy rain prediction. *Chin. J. Atmos. Sci.* **2005**, *5*, 717–726.

20. Stensrud, D.J.; Yussouf, N. Reliable probabilistic quantitative precipitation forecasts from a short-range ensemble forecasting system. *Wea. Forecast.* **2007**, *22*, 3–17.

21. Chen, D.; Shen, X. Recent Progress on GRAPES Research and Application. *J. Appl. Meteorol. Sci.* **2006**, *17*, 773–777.

22. Wang, X.; Bishop, C.H. A comparison of breeding and ensemble transform Kalman filter ensemble forecast schemes. *J. Atmos. Sci.* **2003**, *60*, 1140–1158.

23. Weichert, A.; Burger, G. Linear *versus* nonlinear techniques in downscaling. *Clim. Res.* **1998**, *10*, 83–93.

24. Cannon, A.J.; Whitfield, P.H. Downscaling recent stream-flow conditions in British Columbia, Canada using ensemble neural networks. *J. Hydro.* **2002**, *259*, 136–151.

25. Kuhnlein, C.; Keil, C.; Craig, G.C.; Gebhardt, C. The impact of downscaled initial condition perturbations on convective-scale ensemble forecasts of precipitation. *Quart. J. R. Meteor. Soc.* **2014**, *140*, 1552–1562.

26. Toth, Z.; Kalnay, E. Ensemble forecasting at NCEP and the breeding method. *Mon. Wea. Rev.* **1997**, *125*, 3297–3319.

27. Denis, B.; Cote, J.; Laprise, R. Spectral decomposition of two-dimensional atmospheric fields on limited-area domainsusing discrete cosine transform (DCT). *Mon. Wea. Rev.* **2002**, *130*, 1812–1829.

28. Palmer, T.N.; Gelaro, R.; Barkmeijer, J.; Buizza, R. Singular vectors, metrics, and adaptive observations. *J. Atmos. Sci.* **1998**, *55*, 633–653.

29. Toth, Z.; Zhu, Y.; Marchok, T. The use of ensembles to identify forecasts with small and large uncertainty. *Wea. Forecast.* **2001**, *16*, 436–477.

30. Zhu, Y.; Toth, Z.; Wobus, R.; Richardson, D.; Mylne, K. The economic value of ensemble-based weather forecasts. *Bull. Am. Meteor. Soc.* **2002**, *83*, 73–83.

31. Hersbach, H. Decomposition of the continuous ranked probability score for ensemble prediction systems. *Wea. Forecast.* **2000**, *15*, 559–570.

32. Hamill, T.M. Interpretation of rank histograms for verifying ensembles. *Mon. Wea. Rev.* **2001**, *129*, 550–560.

33. Caron, J.F. Mismatching perturbations at the lateral boundaries in limited-area ensemble forecasting: A case study. *Mon. Wea. Rev.* **2013**, *141*, 356–374.

34. Wang, Y.; Bellus, M.; Geleyn, J.F.; Ma, X.; Tian, W.; Weidle, F. A new method for generating initial condition perturbations in a regional ensemble prediction system: Blending. *Mon. Wea. Rev.* **2014**, *142*, 2043–2059.

7

A Modelling Study of the Impact of On-Road Diesel Emissions on Arctic Black Carbon and Solar Radiation Transfer

Giovanni Pitari *, Glauco Di Genova and Natalia De Luca

Department of Physical and Chemical Sciences, Università degli Studi dell'Aquila, Via Vetoio, Coppito, 67100 L'Aquila, Italy; E-Mails: glauco.digenova@aquila.infn.it (G.D.G.); natalia.deluca@aquila.infn.it (N.D.L.)

* Author to whom correspondence should be addressed; E-Mail: gianni.pitari@aquila.infn.it

Academic Editor: Ivar S.A. Isaksen

Abstract: Market strategies have greatly incentivized the use of diesel engines for land transportation. These engines are responsible for a large fraction of black carbon (BC) emissions in the extra-tropical Northern Hemisphere, with significant effects on both air quality and global climate. In addition to direct radiative forcing, planetary-scale transport of BC to the Arctic region may significantly impact the surface albedo of this region through wet and dry deposition on ice and snow. A sensitivity study is made with the University of L'Aquila climate-chemistry-aerosol model by eliminating on-road diesel emissions of BC (which represent approximately 50% of BC emissions from land transportation). According to the model and using emission scenarios for the year 2000, this would imply an average change in tropopause direct radiative forcing (RF) of -0.054 W·m^{-2} (globally) and -0.074 W·m^{-2} over the Arctic region, with a peak of -0.22 W·m^{-2} during Arctic springtime months. These RF values increase to -0.064, -0.16 and -0.50 W·m^{-2}, respectively, when also taking into account the BC snow-albedo forcing. The calculated BC optical thickness decrease (at $\lambda = 0.55$ μm) is 0.48×10^{-3} (globally) and 0.74×10^{-3} over the Arctic (*i.e.*, 10.5% and 16.5%, respectively), with a peak of 1.3×10^{-3} during the Arctic springtime.

Keywords: black carbon aerosols; global-scale aerosol model; large-scale atmospheric transport; radiative forcing; snow-albedo forcing

1. Introduction

Motor vehicles used for land transportation (automobiles, trains, freight traffic, agriculture) contribute in a significant way to emissions of atmospheric pollutants that are relevant for global climate and/or local air quality, namely CO_2, CH_4, NMHC (non-methane hydrocarbons), CO, NO_x (and O_3 as a photo-chemically-produced secondary species), SO_2 and PM (*i.e.*, aerosol particles, mainly carbonaceous and sulfate) [1]. Quantifying the climate impact of these species in terms of radiative forcing (RF) is a well-assessed exercise only for well-mixed long-lived species (CO_2 and CH_4), whereas it is a much more complex problem (often with uncertain results) for greenhouse gases with shorter lifetimes that may have large spatial gradients (e.g., O_3) and for aerosols that may have highly variable sizes, compositions and spatial distributions [2,3].

Aerosols of anthropogenic origin (mostly from fossil fuel combustion) are normally purely scattering particles (*i.e.*, unity single scattering albedo and negligible absorption of incoming solar radiation), as, for example, sulfate and organic carbon (OC) particles. By increasing the equivalent surface-atmosphere albedo, these aerosols tend to cool the Earth's surface, both directly through solar radiation scattering and indirectly by acting as cloud condensation nuclei. A very different type of particle produced during combustion is that made of elemental carbon, the so-called black carbon aerosols (BC), whose imaginary part of the refractive index is quite large, making the solar radiation absorption by these particles dominant over scattering, with single scattering albedo values for freshly-emitted anthropogenic BC close to 0.2–0.3 at $\lambda = 0.55$ µm [4]. Some recent studies suggest that also OC aerosols (at least from biomass burning sources) can be absorbing, due to their partial BC content [5]. Most BC particles have a relatively short lifetime (approximately one week), but in some recent studies, it was suggested that BC particles are the second-most important climate-forcing agent after CO_2 [6,7].

The interest in anthropogenic BC aerosols, beyond their role in offsetting the cooling due to sulfate, mineral dust, sea salt and organic aerosols, also regards their pollutant aspect. Several studies have demonstrated the negative health impact of fossil fuel combustion BC from inhalation, due to their small size and consequent ability to penetrate deep into the respiratory system, promoting allergy and cancers [8]. Neurodegenerative effects on children have also been shown [9], as well as a potential reduction of the length of telomeres, a measure of biological ageing, thus increasing the risk of atherosclerosis, diabetes, hypertension, coronary artery disease and heart failure [10].

Open biomass burning (*i.e.*, grass and forest fires) is another important source of BC: in this case, there is normally a large burden of co-emitted organic material that can quickly condense onto the BC cores, thereby increasing the single scattering albedo (with canonical values at $\lambda = 0.55$ µm close to 0.7–0.8). Among anthropogenic emissions, diesel engines used for land transportation are the most important BC source. Diesel engines are very common in developing countries, East Asia and Eastern Europe and also in Western Europe and the U.S., because market strategies and policy decisions have encouraged these engines compared to gasoline engines, because they are more fuel-efficient, thus emitting less CO_2/km. Despite their higher fuel-efficiency, diesel engines release more particles, including BC [11,12]. These engines account for approximately one-third of the global anthropogenic emissions of BC and approximately 17% of the total [13]. The relative weight of BC diesel emissions may be substantially larger on a regional basis, mainly in western industrialized countries. In the U.S., for example, diesel engines represent almost 27% of the total BC emissions [14].

Because of the large warming effect per unit mass of BC emitted, the reduction of diesel engines is a good candidate to mitigate climate forcing [15]. After a large expansion of diesel engines in public transportation, recent concerns about the environmental impact of these engines has encouraged some municipal transportation agencies around the world to look for replacement technologies or at least complementary alternatives to diesel buses. In 2013, the Los Angeles Metro switched the entire bus fleet to compressed natural gas; the same policy was adopted in New Delhi (India) for all public vehicles in 2001. In New York City, after a period of operating hybrid-electric buses, they readopted diesel buses in 2013, due to poor hybrid engine performance.

Non-CO_2 emissions due to land transportation are estimated to have a significant climate impact, with a net RF of approximately 20% of total anthropogenic CO_2 emissions [1]. The WMO [16] assessment study has calculated that a large reduction of BC emissions coupled with reductions of NO_x, NMHC and CH_4 may reduce the radiative forcing by up to 50% globally and by 2/3 over the Arctic region, if combined with CO_2 emission cuts. In addition, the beneficial effects of BC reduction could be achieved in a few weeks, due to its short atmospheric lifetime. New technologies are available for diesel engines with the use of innovative and more efficient anti-particulate filters (DFP) that may reach a 90% BC cut-off from the exhaust, requiring, however, the use of fuel with very low sulfur content (ULSD).

One important aspect of the potential emission reduction of pollutants in Northern Hemisphere source regions is that these short-lived species may be effectively transported over large spatial scales with significant impact on remote regions. As discussed in Pitari et al. [17], the local concentration of aerosols in an urban site may be greatly perturbed by large-scale transport events of desert dust and forest fire smoke. Similarly, remote sites may be impacted by planetary-scale transport of aerosols from anthropogenic sources, namely pollutants from motor vehicle emissions and other sources. BC transport from Northern Hemisphere (NH) mid-latitudes over the Arctic region is an important example of this source-receptor problem [18].

Although the Arctic climate has a complex meteorological system, its major features could be summarized as follows: (1) the formation of a strong polar vortex during winter with relatively stable stratification in the troposphere; (2) weakening of the polar vortex in late winter and spring months, allowing greater exchange of low level with upper level air, due to breaking of the vertical stability. Upper level air, in turn, is more efficiently affected by transport from the mid-latitudes due to more intense zonal and southerly winds. BC particles transported to the Arctic polar latitudes and deposited over snow and ice on the surface [19] may have important consequences on the polar surface albedo and on local and global climate forcing [20,21]. In addition, the solar radiation absorbing efficiency of BC particles located above a high albedo surface (i.e., ice and snow) is by far larger than for other types of soils or the ocean [22].

The focus of the present study is the quantification of the potential mitigation effect due to a reduction of BC emissions from land traffic to be (realistically) obtained by a complete reconversion of on-road diesel engines to other forms of vehicle propulsion (i.e., gasoline, electric, etc.) or with new technologies capable of controlling soot emissions. Here, a simple "upper-limit" approach is adopted by setting all on-road diesel emissions of BC to zero and leaving all other relevant species emissions (i.e., NO_x, CO, OC, sulfate, off-road BC) unchanged. Detailed radiative transfer calculations are carried out over the global domain with a specific focus on the Arctic region, including the BC snow-albedo forcing.

2. Experimental Section

A brief description of the climate-chemistry-aerosol model used in this study and its basic setups are presented in the following subsections, along with a description of the adopted emission inventories and overall presentation of the numerical experiment setup.

2.1. The Model

The University of L'Aquila model used in this study is a global-scale climate-chemistry coupled model with an interactive aerosol module (ULAQ-CCM). This model has been fully described in Pitari *et al.* [23] and also in Eyring *et al.* [24] and Morgenstern *et al.* [25] for the climate-chemistry model validation initiative, SPARC-CCMVal (Stratospheric Processes and Their Role in Climate-Climate Chemistry Model Validation). Since then, some important updates have been made to the model: (1) increases in horizontal and vertical resolution, now T21 (5° × 6° lat × lon) with 126 log-pressure levels (approximate pressure altitude increment of 568 m); (2) inclusion of a numerical code for the formation of upper tropospheric cirrus cloud ice particles [26,27]; (3) upgrade of the radiative transfer code for calculations of photolysis, solar heating rates and radiative forcing. This updated model version was used and documented in Pitari *et al.* [28].

The ULAQ model includes the major components of stratospheric and tropospheric aerosols (sulfate, carbonaceous, soil dust, sea salt), with the calculation at each size bin of surface fluxes, removal and transport terms, in external mixing conditions. BC and OC particles are treated separately; the BC fraction emitted by fossil fuel combustion is assumed to be hydrophobic [29], with an ageing time of 1 day [30] (although this time may vary significantly, depending upon co-emitted species). BC is treated in a bin model, with 7 bins from 10 nm to 0.64 μm by doubling the particle radius and assuming a log-normal distribution of the emissions. BC from fossil fuel and biomass burning sources are treated separately. A $0.1\ cm \cdot s^{-1}$ surface dry deposition velocity is used for BC ($0.03\ cm \cdot s^{-1}$ on snow/ice). Wet deposition removal is calculated as the product of climatological precipitation rates and a scavenging coefficient ($2.1\ cm^{-1}$ for stratiform precipitation and $0.6\ cm^{-1}$ for convective precipitation), using a climatological cloud distribution. BC, OC and SO_2 surface fluxes are those made available for the climate-chemistry model initiative (CCMI) community [31] and relative to the year 2000. Details on the treatment of emissions and removal processes (*i.e.*, wet/dry deposition and gravitational settling) can be found in Textor *et al.* [32] and Kinne *et al.* [33]. The ULAQ model has extensively participated in several aerosol evaluation campaigns [34–37] and in BC-specific studies under the AeroCom project [19,30,38] (Aerosol Comparisons between Observations and Models).

Since the first radiative calculations made with the ULAQ model in the framework of AeroCom Phase I [22], a new radiative transfer module has been included. It is a two-stream delta-Eddington approximation model operating on-line in the ULAQ-CCM and used for chemical species photolysis rate calculation at UV-visible wavelengths and for solar heating rates and radiative forcing at UV-VIS-NIR bands [39]. In addition, a companion broadband, k-distribution longwave radiative module is used to compute radiative transfer and heating rates in the planetary infrared spectrum [40], with stratospheric temperature adjustment for the calculation of top-of-atmosphere/tropopause RFs. Aerosol optical parameters are calculated with a Mie scattering program [41], using wavelength-dependent refractive indices as the input mainly from the OPAC database and the aerosol size distribution from the ULAQ model. Calculations of photolysis rates and radiative fluxes have been validated in the framework of SPARC-CCMVal [42] and AeroCom inter-comparison campaigns [39]. In the latter case, the radiative transfer code was validated for the AeroCom Phase II direct aerosol effect (DAE) experiment, where 15 detailed global aerosol models have been used to simulate the changes in the aerosol distribution over the industrial era.

2.2. BC Radiative Properties

Carbonaceous particles are both primary (BC and OC) and secondary particles (OC), the latter mostly originating from emissions of complex hydrocarbons from vegetation (terpene families) and leading to the formation of the so-called secondary organic aerosols (SOA). Anthropogenic combustion produces primary carbonaceous particles with an organic fraction (OC) or pure elemental carbon (BC). BC particles have a large hydrophobic fraction, with a bulk density of approximately 1.7 $g \cdot cm^{-3}$ [6] and a size distribution in the accumulation mode with a mean radius close to 0.07–0.1 µm (see, for example, Schwarz et al. [43] and Pueschel and Kinne [44]). Sulfate (SO_4) and OC, on the other hand, are highly soluble particles.

The most important difference among these aerosol families, however, is in the optical characteristics of the particles, with SO_4 and OC acting as almost purely scattering aerosols in the visible range of the solar spectrum. BC, in contrast, has a significant imaginary part of the refractive index and a single scattering albedo in the visible range close to 0.4–0.5 when an organic material coating of ~50 nm in thickness is considered [43]. These characteristics imply a strong absorbing capacity of BC aerosols, which produces a warming in the vertical layer containing BC particles. If the surface albedo is sufficiently high, the resulting radiative forcing will be positive, due to the BC optical depth, and have a high normalized value per unit optical depth. OC and SO_4, in turn, tend to cool the surface both directly, by increasing the atmospheric reflectivity, and indirectly, by acting as cloud condensation nuclei. More cloud droplets, in turn, may increase the equivalent surface-atmosphere albedo. The aerosol-cloud indirect effect, however, is not considered in the present study.

In a remote site (such as the Arctic region), the aerosol concentration is determined by spatial-temporal features of planetary-scale transport and by the efficiency of anthropogenic emissions in the polluted source regions, coupled to local removal processes. Consequently, a meaningful impact study on aerosols located in a remote region requires good knowledge of the anthropogenic emissions in the source regions (1), their potential perturbations related to trends or pollution-control measures (such as a limitation of on-road diesel engines) (2), the efficiency of irreversible loss rates (3) and a proper description of the large-scale transport pathways towards the receptor region (4). The Arctic region is vulnerable to BC transport from Northern Hemisphere mid-latitude source regions; over the Arctic, not only is the radiative efficiency of BC the highest, but its snow-albedo forcing is also potentially important [19].

2.3. BC Emissions

According to WMO [16] (World Meteorological Organization), open biomass burning dominates BC emissions on a global scale (41%), with the largest contribution from the tropics. Total land transportation accounts for 19% of the total and 37% of the anthropogenic fraction; diesel on-road transportation emissions account for 10% of the total and 19% of the anthropogenic fraction. Figure 1 is adapted from Bond *et al.* [13] and shows the on-road diesel percent fraction of total land transportation BC emissions on a regional basis and on-road diesel and total BC emissions in absolute units. The relative contribution of BC emissions from on-road diesel is close to 50% of the total land transport BC emissions in western industrialized countries. The relative contribution of on-road diesel to total BC emissions ranges from 2% in China to 29% in Europe and 48% in the Middle East. As mentioned above, gridded emissions for the year 2000 are taken from the CCMI database [31,32].

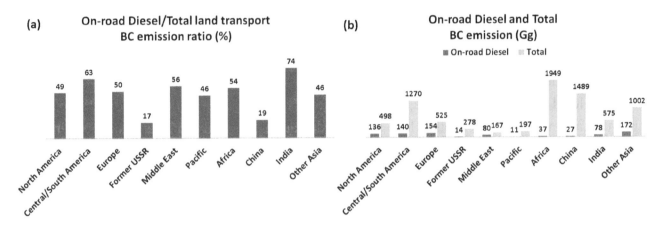

Figure 1. (**a**) On-road diesel percent fraction of total land transport black carbon (BC) emissions (regional). (**b**) On-road diesel and total BC emissions (Gg/yr) (regional). Adapted from Bond *et al.* [13].

2.4. Numerical Experiment Setup

This study is organized through a baseline experiment ("base") with anthropogenic fossil fuel emissions of BC, OC and SO$_2$ from the year 2000 and with (climatological) biomass burning sources and a perturbed experiment ("pert"), where on-road diesel emissions of BC particles are removed from the total land transportation sources, using the regional fractions of Figure 1a. These fractions represent a contribution of 27% and 29% of total BC emissions over North America and Europe, respectively, and 14% and 17% over India and East Asia, respectively, excluding China (2%), where BC emissions from on-road diesel are relatively low compared to domestic fire or other non-transportation burning sources. Both "base" and "pert" simulations are run for 6 model years (2000–2005) after a 2-year spin-up (1998–1999); the results for aerosol and gas species are then averaged over 2000–2005 to allow for more robust calculations of anomalies (*i.e.*, "pert"-"base"). The sensitivity approach outlined in this work should be intended as an upper limit of the feasible mitigation, because no displacement of other engine propulsion forms is considered here. In addition to these two main experiments, five additional sensitivity studies were carried out to quantify the model transport of BC from source regions to the Arctic (see the summary in Table 1 and the discussion on BC aircraft emissions).

Table 1. Percent contribution of Northern Hemisphere regional emissions to the Arctic optical thickness of BC (65°N–90°N). Annual average.

	$\Delta\tau$ (%)
Europe	33
Russia	32
Asia	27
North America	8

3. Results and Discussion

The model calculated global BC burden is 0.27 Tg C with an average lifetime of 7.2 d; BC global removal is approximately 65% by wet deposition and 35% by dry deposition (77% and 23%, respectively, over the Arctic region, 65°N–90°N), consistent with global budgets reported in Koch and Hansen [18]. Global AeroCom statistics reported in Textor *et al.* [32] give a mean value of 79% for BC wet deposition among six models, with a 10% standard deviation. Detailed results are discussed in the following subsections, starting from an evaluation of the aerosol results with available observations in the Arctic region. The impact of the removal of on-road diesel emissions is then discussed, both on the calculated BC optical thickness and on the radiation budget.

3.1. Evaluation of Model Results with Observations

A comparison of model results for the surface mass concentration of BC with observations on selected worldwide measurement sites is presented in Figure 2a, with the thin lines highlighting a factor of two deviations. The observations used here are those referenced by Liousse *et al.* [45] and Penner *et al.* [34]. A similar evaluation of the model results is made for the Arctic region in Figure 2b, using monthly averaged values over three Arctic stations (Barrow, Alert and Ny Alesund). In this case, the observed values are decadal averages reported in Quinn *et al.* [46] and Sharma *et al.* [47]. A model overestimation is observed at Barrow during the BC peak concentration time of the year (*i.e.*, springtime) and a systematic positive bias for the summertime minimum values at all stations (see Figure 3).

Figure 3a is an example of the model calculated BC mass mixing ratio vertical profile in the Arctic Pacific region compared to aircraft measurements made during the NOAA ARCPAC campaign during April, 2008 [43]. The model features a maximum in the vertical profile at approximately 3.5 km altitude close to 150 ng/kg, consistent with observations, although the mixing ratio decline above this height is less evident in the observations. The ARCPAC measured increase in BC concentration with altitude is consistent with the aircraft observations of Rosen *et al.* [48]: they reported that BC in aircraft observations over Barrow increases by a factor of three above the boundary layer. The ULAQ model calculates an increase by a factor 1.8 in the BC mass mixing ratio at 3.5 km altitude with respect to the surface. However, the BC vertical profile predicted in the ULAQ model is in the uncertainty bar of the measurements at all heights and is improved with respect to previous calculations [38], mostly due to a better representation of wet deposition. Recent AeroCom Phase II studies [49], targeted at global BC circulation modelling, clearly state the need for more extensive flight measurement campaigns to properly characterize the long-range BC transport and its atmospheric lifetime.

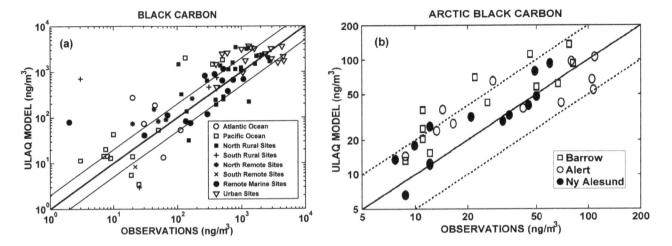

Figure 2. (**a**) Scatter plot of calculated *versus* observed annual mean BC mass concentration values (ng/m^3) at selected worldwide locations (see the text and legend). (**b**) Same as in (a), but for monthly averaged values at three selected Arctic stations (see the text and legend). Thin (or dashed) lines highlight a factor of two deviations.

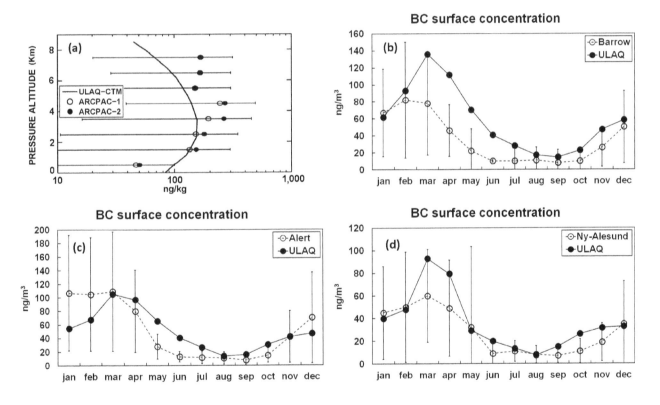

Figure 3. (**a**) Vertical profile of the model calculated BC mass mixing ratio (ng/kg) compared to airborne measurements taken during the ARCPAC campaign (April, 2008, Arctic Pacific flights) [43]. (**b–d**) Seasonal cycle of surface mass density (ng/m^3) from the model, compared to observations at selected Arctic stations: Barrow (71°N, 156.6°W), Alert (82°N, 62.3°W) and Ny-Alesund (79°N, 12°E), respectively [46,47].

Figure 3b–d compares the modelled BC seasonal cycle over the three Arctic stations in Figure 2b with decadal averages from available observations [46,47]. Filter-based absorption instruments at these Arctic stations include the Particle Soot Absorption Photometer (PSAP) and the Aethalometer. The modelled BC concentration seasonal changes appear to be quite similar to the findings from the ACCMIP model inter-comparison [50] (The Atmospheric Chemistry and Climate Model Intercomparison Project). For the three considered measurement sites (Barrow, Alert and Ny-Alesund), northern Eurasia has been identified as the major source region, especially in winter/spring. The ULAQ model is consistent at all three stations, except for a significant overestimation at Barrow during springtime months and some overestimation of the summer minimum at all stations. As discussed in the report of Quinn *et al.* [45], the models' ability to reproduce observed measurements can be heavily impaired by the treatment of BC microphysical properties and removal rates. A general behavior is shown by several models discussed in the report to underestimate winter/spring and to overestimate summer/fall BC concentrations in the Arctic. The ULAQ model overestimation during fall months is in the uncertainty interval of the recorded observations, whereas springtime BC concentrations are overestimated, particularly at Barrow.

3.2. Discussion of Model Results

Mie scattering properties relevant to the modelled BC size distribution have been calculated through a standard Mie scattering code [41] to provide the wavelength-dependent optical properties needed for a radiation transfer analysis and for calculating the BC optical thickness from the mass density distribution. Several BC aerosol size distribution measurement campaigns provide a quite accurate estimate of the lognormal distribution parameters for transported BC aerosols [43,44]. Accordingly, an effective radius of 0.14 μm has been selected as the input to the Mie code (0.05 μm of which are due to coating thickness) [43,51]. The BC complex refraction index is taken from Bond *et al.* (2013) [6]. A single scattering albedo of 0.45 is calculated at $\lambda = 0.55$ μm, with a refraction index of 1.95-0.79i at the same wavelength. For biomass burning BC, the single scattering albedo is increased to 0.75.

The model-calculated optical thickness of black carbon aerosols and its seasonal evolution in the Arctic region are presented in Figure 4. Consistent with other model calculations [18], the larger values on the eastern Arctic and the shape of the contour lines highlight the important role of the large-scale transport of BC aerosols from Eurasian sources. Table 1 presents a summary on an annual basis of the model calculated continental contributions to Arctic BC.

According to the model, Europe and Russia together (including both industrial sources and biomass burning) provide the major input to the Arctic BC optical thickness. These sources, along with those from Southeast Asia, also provide substantial input to the Western Arctic BC and over the Atlantic, through coupling of the subarctic Westerlies and northward eddy mass fluxes in the atmospheric layers above the surface. The contribution of aircraft emitted BC is negligible in terms of total optical depth (<1%), but larger in the Arctic upper troposphere and lower stratosphere (UTLS): the ULAQ model calculates an aircraft contribution of 4.5% on an annual basis, using the base case emission scenarios adopted from the EC-REACT4C project [52] with an average soot emission index of 0.013 g/kg-fuel at cruise altitudes. Although the UTLS direct effect of BC aircraft emissions is small, their potential climate impact could be significant through indirect formation of "soot-cirrus" particles when ice super-saturation conditions are found, and ice particle formation may occur via heterogeneous freezing [53].

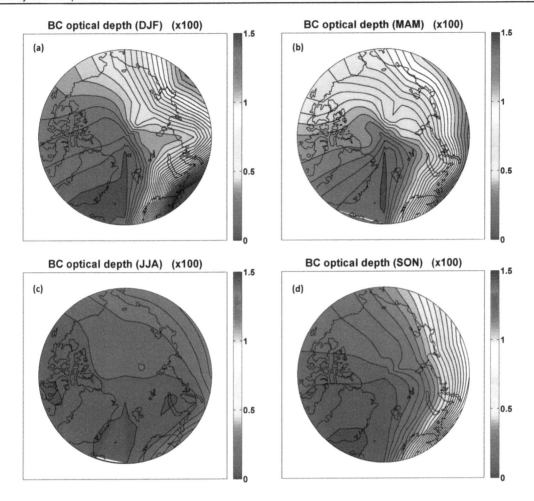

Figure 4. Model calculated BC optical thickness (×100) at λ = 0.55 μm in the Arctic region: (**a–d**) from top left to bottom right, for December-January-February (DJF), March-April-May (MAM), June-July-August (JJA), September-October-November (SON) averages, respectively. The contour line step of optical thickness (×100) is 0.05, starting from 0.15, 0.20, 0.05 and 0.10 in (a–d), respectively, with values of minimum optical depth located on the NW Atlantic.

The pronounced seasonal cycle of Arctic BC optical depth is evident looking at the 65°N–90°N monthly averaged values in Figure 5a,b, where the seasonal cycle per geographical sector is presented. Values are generally between 0.003 and 0.008 throughout the Arctic during winter and spring, except over Greenland and the northwest Atlantic, where it is less (see Figures 4 and 5b), and over the Eastern Arctic during February and March, where values are between 0.012 and 0.014. Optical depth values smaller than 0.003 are predicted during summer and fall seasons when subarctic Westerlies and northward transport due to eddy mass fluxes and mean meridional winds are less pronounced than during winter and spring months. Maximum BC transport is predicted in the model in the late winter and early spring, during Arctic haze transport events, when the greatest transport of pollutants occurs in the boundary layer [54]; a peak of the calculated BC optical thickness is visible during March (0.004) in the lowest 2 km altitude (Figure 5c). The modelled optical thickness during February-March-April is a factor of two higher with respect to the annual average, and the seasonality (*i.e.*, winter-summer difference) is stronger over the eastern Arctic compared to the western Arctic.

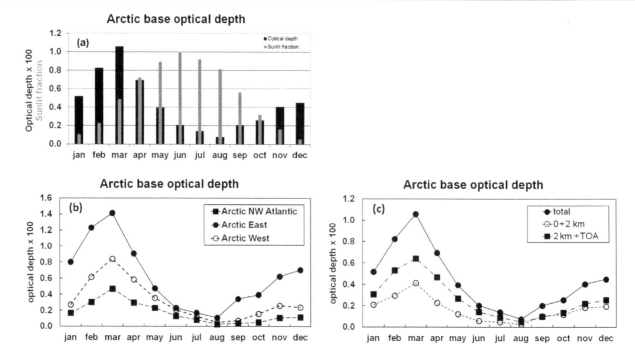

Figure 5. (**a**) Monthly averaged values of model calculated BC optical thickness at λ = 0.55 μm in the Arctic region (65°N–90°N), with the sunlight fractional coverage superimposed. (**b**) As in (a), but over longitude ranges 60°W–10°E, 10°E–180°E and 180°E–60°W. (**c**) As in (a), but also for altitude layers 0–2 km and 2 km up to the top of the atmosphere (TOA).

The insights of the model large-scale transport towards the Arctic can be obtained from Figure 6a; in this way, the modelled optical thickness seasonal behavior reported in Figure 5b,c can be better explained, as well as the time series of surface BC mass concentrations in Figure 3b–d. The mean meridional BC mass flux [vχρ] at 65°N (positive northward, *i.e.*, the Arctic influx) is shown for the free troposphere and for the boundary layer (up to 850 hPa). In the latter case, the mean meridional circulation component of the flux [v][χρ] is also presented ([] indicates a zonal average). The dominant role of eddy mass fluxes {[v'(χρ)'] = [vχρ]-[v][χρ]} is evident during the Arctic haze period close to the surface, as reported in the real world (Barrie, 1986). The most favorable conditions for efficient northward transport are found in the model over the Atlantic/European region and in the North American Pacific (Figure 6b), consistent with observations [55]. Hansen and Nazarenko [21] note that these observed features of BC transport towards the Arctic are produced by meteorological blocking conditions during winter-spring months, driving the circulation around the Icelandic low and Siberian high.

Model-calculated changes in the Arctic averaged optical thickness due to the cancellation of on-road diesel sources are shown in Figure 7a (*i.e.*, "pert"-"base" differences) and closely follow the same seasonal cycle of the baseline reference optical depth shown in Figure 5a, with a rather flat relative decrease during the year (16.5% ± 2%) (Figure 7b). This is indirect evidence that the greatest amount of BC loading over the Arctic is transported from NH mid-latitude sources and may therefore be significantly affected by changes in anthropogenic fossil fuel burning, both in terms of decadal trends and the potential regulating measures of these emissions. On a global scale, the BC optical thickness decrease in the sensitivity experiment is calculated to be 10.5% ± 2% (Figure 7c). The decrease in the relative change in BC optical thickness during summer months (8% in July–August *versus* 13% in April) results primarily from the impact of biomass burning sources in Russia during these months [56]. The contribution of boreal forest fires may actually be highly variable from year to year due to different meteorological conditions. The ULAQ model does not consider this interannual variability, because it uses a climatological inventory dataset for open biomass burning emissions.

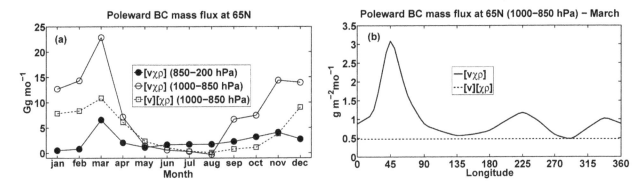

Figure 6. (**a**) Time series of the mean zonal poleward BC mass flux at 65°N latitude (Gg/month), calculated as [vχρ], where [] indicates a zonal average, χ is the BC mass mixing ratio, ρ is the atmospheric mass density and v the meridional wind. The dashed line is for the boundary layer mass flux (1000–850 hPa) due to the mean meridional circulation only {[v][χρ]}. (**b**) March average of the poleward BC mass flux at 65°N latitude (g·m^{-2}·mo^{-1}) in the boundary layer (1000–850 hPa) as a function of longitude {[vχρ]}. The dashed line is the flux due to the mean meridional wind only; [] indicates a zonal average.

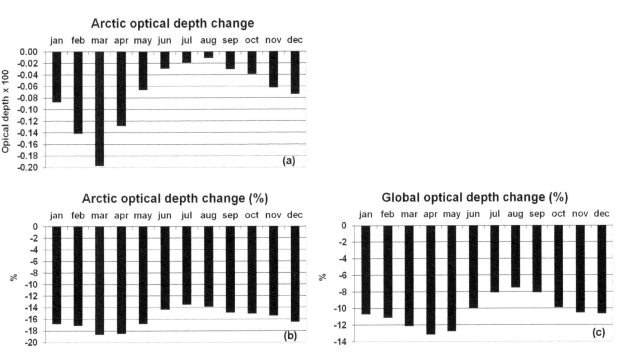

Figure 7. (**a**) Monthly averaged values of model calculated BC optical thickness changes at λ = 0.55 µm averaged over the Arctic region (65°N–90°N), with changes due to the cancellation of on-road diesel emissions of BC with respect to the base case ("pert"-"base"; pert, perturbed). (**b**) As in (a), but for percent changes. (**c**) Monthly-averaged global percent changes.

3.3. Radiative Calculations

Except when specified, RF values discussed in this section are calculated in total sky conditions at the tropopause. The Arctic regional radiative forcing associated with the BC perturbation described above is shown in Figure 8a, whereas Figure 8b refers to the globally-averaged RF; here, the direct radiative impact is shown along with the BC snow-albedo forcing. The springtime peak in Arctic BC RF (Figure 8a) is due to coupling of the maximum optical depth change and the incoming solar radiation increase with respect to the winter months (see Figure 5a). As pointed out in Flanner [20], April–May tropospheric BC induces the greatest normalized radiative impact in the Arctic, because high insolation and surface albedo help reach a large specific forcing. According to the model, the average direct RF in the Arctic during March-April-May amounts to -0.22 W·m^{-2}, to be compared with an annual global average of -0.054 W·m^{-2} and an annual Arctic average of -0.074 W·m^{-2} (see Table 2). Including the BC snow-albedo forcing, these RF values increase to -0.50, -0.064, -0.16 W·m^{-2}, respectively.

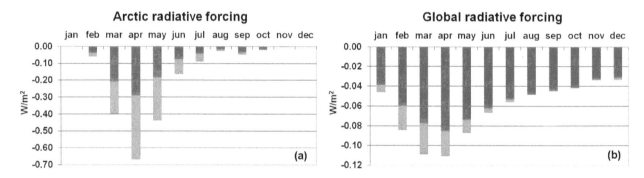

Figure 8. Monthly averaged values of tropopause radiative forcing (W·m^{-2}) due to model calculated BC optical depth changes from the cancellation of on-road diesel emissions of BC with respect to the base case ("pert"-"base"). (**a,b**) Arctic and global averages, respectively. Red/blue bars are for direct and snow-albedo radiative forcings (RFs), respectively.

Table 2. BC optical thickness "pert-base" changes at $\lambda = 0.55$ µm, related radiative forcing (W·m^{-2}) and normalized radiative forcing (NRF) (W/m^2/$\Delta\tau$) averaged over the Arctic region (65°N–90°N) and globally. NRF is computed on a monthly basis, then the yearly average is taken. The fifth column shows the BC mixing ratio "pert-base" changes in Arctic snow due to wet and dry deposition.

	$\Delta\tau$ ($\times$100)	Direct RF (W·m^{-2})	Direct NRF (W/m^2/$\Delta\tau$)	Snow-Albedo RF (W·m^{-2})	BC in Arctic Snow (ppbm)	Total RF (W·m^{-2})	Total NRF (W/m^2/$\Delta\tau$)
Arctic annual	−0.074	−0.074	110	−0.086	6.0 ± 3.6	−0.16	240
Global annual	−0.048	−0.054	115	−0.010		−0.064	130
Arctic MAM	−0.13	−0.22	200	−0.28	10.5 ± 5.5	−0.50	460
Global MAM	−0.055	−0.079	145	−0.023		−0.10	185

A potential overestimate of the calculated spring-summer RF values may be expected when considering the ULAQ model tendency to overestimate surface BC values at some Arctic locations (see, for example, Figure 3b,c). Note also how the seasonal cycle of the global RF (Figure 8b) is similar to the Arctic RF, indicating the significant radiative impact of poleward-transported BC. The BC direct radiative effect in the ULAQ model exhibits a clear dependence on its vertical distribution, with a significantly higher forcing efficiency at altitudes above 5 km (see Table 3), consistent with the findings of Samset et al. [57]. As in our case, they show an efficiency increase with increasing altitude and a decrease when approaching the tropopause.

Table 3. Calculated BC radiative forcing efficiency $(W/m^2/\Delta\tau)$ as a function of altitude, for clear sky and total sky conditions (Arctic MAM).

Pressure Layer (hPa)	Direct NRF Clear Sky $(W/m^2/\Delta\tau)$	Direct NRF Total Sky $(W/m^2/\Delta\tau)$
334–284	87	99
393–334	207	220
463–393	203	215
544–463	198	210
640–544	193	203
753–640	187	184
885–753	182	171
1000–885	176	164

Snow/ice albedo changes due to BC deposition are not explicitly calculated in this study. However, the annually averaged BC deposition over Arctic snow/ice resulting from the baseline model simulation implies a BC mass mixing ratio in surface H_2O of 36 ± 22 ppbm (see Table 4), with uncertainty related to the geographical dispersion, and this value is consistent with that reported by Hansen and Nazarenko [21] (i.e., 30 ppbm mean value and 4 ppbm over Greenland, where minimum BC deposition is recorded). For this reason, their same albedo perturbation is applied here (i.e., 1.5% on Arctic and 3% on snow-covered northern mid-latitudes), although substantial uncertainty must be admitted, because the actual albedo perturbation is a function of important microphysical factors (in particular, the internal/external mixing of BC and fresh/old snow conditions). The 1.5% relative change in snow/ice albedo over the Arctic region has to be considered a spatially, annually and spectrally averaged value, with a 2.5% change imposed for $\lambda < 0.77$ μm, 1.25% for 0.77 μm $< \lambda < 0.86$ μm and no change for $\lambda > 0.86$ μm (as in Hansen and Nazarenko [21]).

Table 4. Base case optical thickness of BC at $\lambda = 0.55$ μm and related radiative forcing $(W \cdot m^{-2})$. The fourth column shows the BC mixing ratio in Arctic snow due to wet and dry deposition.

	τ ($\times 100$)	Direct RF $(W \cdot m^{-2})$	Snow-Albedo RF $(W \cdot m^{-2})$	BC in Arctic Snow (ppbm)	Total RF $(W \cdot m^{-2})$
Arctic annual	0.45	0.43	0.51	36 ± 22	0.94
Global annual	0.46	0.46	0.054		0.51
Arctic MAM	0.72	1.25	1.58	58 ± 31	2.83
Global MAM	0.45	0.66	0.13		0.79

The model calculated deposition fields at each surface grid box are used to scale these albedo perturbation values; the results of these calculations are summarized in Table 4 for total BC and in Table 2 for the BC reduction due to the cancellation of on-road diesel emissions. According to the model, the corresponding BC/snow global-annual climate forcing is +0.054 $W \cdot m^{-2}$ for total BC and −0.010 $W \cdot m^{-2}$ for the BC reduction, with Arctic annual average values of +0.51 $W \cdot m^{-2}$ and −0.09 $W \cdot m^{-2}$, respectively. The albedo perturbation estimated in Hansen and Nazarenko [21] yields a climate forcing of +0.3 $W \cdot m^{-2}$ in the Northern Hemisphere (i.e., ~+0.15 $W \cdot m^{-2}$ globally). In addition, they noted that the efficacy of this indirect soot forcing is ~2, which is twice as effective as CO_2 in altering global surface air temperature for a given radiative forcing. It represents for this reason an important driving mechanism for recent trends in sea/ice temperatures over the Arctic region. Their forcing is then a factor three larger with respect to the one calculated here. However, it was subsequently corrected in Hansen et al. [58] and Hansen et al. [59] to +0.05 $W \cdot m^{-2}$ (globally) by scaling the BC/snow forcing on gridded deposition fields (as in our case) instead of assuming spatially-annually uniform snow albedo reductions over Arctic sea-ice and Northern Hemisphere snow-covered land, as discussed in Flanner et al. [60].

Figure 9a clearly shows how the springtime polar RF contour lines follow the optical depth change contours and highlight transport streamlines from coupled subarctic Westerlies and northward transport. Available observations show high tropospheric BC concentrations over the Arctic up to 8 km in altitude, with magnitudes comparable to profiles measured over polluted areas at northern mid-latitudes [48], thus demonstrating the role of global atmospheric transport mechanisms. As discussed above, the consequent significant BC deposition over Arctic snow and ice tends to reduce the surface albedo [19], thus indirectly contributing to regional (and global) warming. The direct climate effect of Arctic BC is related to the absorption of solar radiation that is efficiently reflected upwards by the snow/ice covered surface, preventing a significant fraction of the surface reflected solar radiation to reach the top of the atmosphere. The latter is a form of radiative process similar to that of greenhouse gases, but acting on solar shortwave radiation instead of on longwave blackbody planetary radiation.

Figure 9b presents model results for the BC snow-albedo RF due to the removal of on-road diesel emissions (spring months): RF contours closely follow the calculated albedo changes, except for some non-linear effects arising from the superposition of spatial inhomogeneity in surface albedo and the distribution of atmospheric absorbers (namely, BC). The total BC RF (direct + snow-albedo effect) arising from the emission perturbation is presented in Figure 9c by summing the contributions in the previous two panels; this value reaches approximately −1.0 $W \cdot m^{-2}$ during spring months over the eastern Arctic region. As discussed above, the snow-albedo BC forcing is calculated on the basis of the perturbation of BC deposition fields (wet + dry). These are shown in Figure 9d and range between approximately 3 and 20 ppbm (i.e., ng-BC/g-H_2O) over Iceland and Siberia, respectively. The snow-albedo BC forcing presented in Figure 9b follows the calculated changes in BC deposition, except over the NE Atlantic and northern Scandinavia, where surface albedo (and its absolute changes due to BC deposition) are much smaller than in all other parts of the Arctic region. On average, annually, the ULAQ model calculates 77% and 23% wet and dry contributions, respectively, to BC Arctic deposition with 0.27 Tg/yr of total deposited BC. As a comparison, Breider et al. [56] calculated 89% and 11% wet and dry contributions, respectively, with 0.28 Tg/y of total deposited BC.

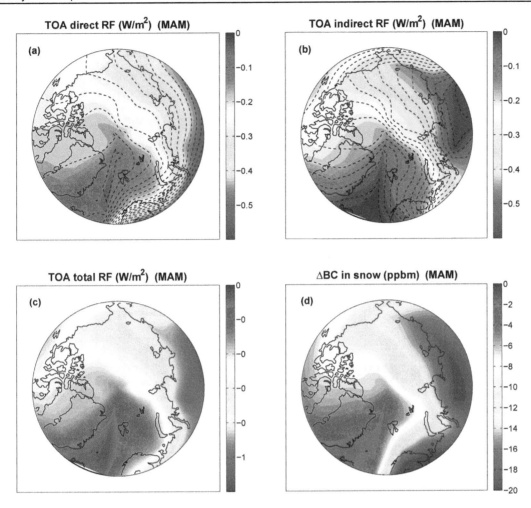

Figure 9. (**a–c**) Model calculated tropopause radiative forcing (W·m^{-2}) defined as in Figure 8, in the Arctic region (March-April-May average) (color scale): direct, snow-albedo and total RF are presented in (a–c), respectively. Superimposed in (a) are the contour lines of the corresponding BC optical depth changes at $\lambda = 0.55$ μm (dashed) and defined as in Figure 7, with a contour line step of -2×10^{-4} starting from -4×10^{-4} on the NW Atlantic, up to -4×10^{-3}. Superimposed in (b) are the contour lines of snow/ice surface albedo absolute changes in the visible wavelength range (dashed), with contour line step -5×10^{-4} starting from -5×10^{-4} on the NW Atlantic, up to -8×10^{-3}. Albedo percent changes are obtained as described in the text, as a linear function of the model calculated changes in BC surface deposition on snow (wet + dry) that are presented in (**d**) in ppbm (*i.e.*, ng-BC/g-H$_2$O).

Figure 10 shows the radiative forcing per unit optical depth both in the Arctic (Figure 10a) and globally (Figure 10b). As discussed above, the BC forcing efficiency in the Arctic region is strongly affected by the underlying surface albedo, substantially higher than its global average due to permanent snow/ice coverage. The normalized radiative forcing (NRF) shown in Figure 10a and summarized in Table 2 follows the seasonality of the available amount of incoming solar radiation and reaches a maximum of approximately 250 in the Arctic during May and June, to be compared with approximately 150 during the same months on a global scale. These values refer only to the direct forcing, for which the definition of NRF is more appropriate. However, because the perturbation in optical thickness also indirectly drives the change in snow/ice albedo, it is possible to include the effects of both direct and snow-albedo RF in the calculation of the normalized forcing. In this case, the ULAQ-model calculated NRF rises up to 650 in the Arctic region during May, and the global value increases to 190 in April (see Figure 10b). On an annual basis, the calculated direct NRF accounts for 110 and 115 over the Arctic and globally, respectively, and increases to 240 and 130, including also the snow-albedo effect. The annual mean NRF reported in the study of Schulz *et al.* [22] for total atmospheric BC is 153 ± 68 on a global scale (direct forcing only). During springtime months with the highest BC RF (see Figure 10), the NRF accounts for 200 and 145 (Arctic and global MAM averages, respectively), increasing to 460 and 185 when considering the snow-albedo effect, as visible in (a) and (b) of Figure 10, respectively. The seasonal behavior of the calculated BC forcing efficacy is consistent with the findings of Flanner [20].

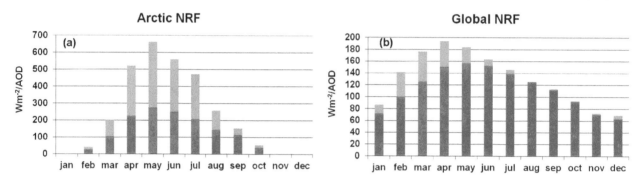

Figure 10. Monthly averaged values of normalized radiative forcing (NRF = RF/$\Delta\tau$) (W·m^{-2}/$\Delta\tau$), where $\Delta\tau$ is the BC aerosol optical depth (AOD) change at $\lambda = 0.55$ μm, defined as in Figure 7. (**a,b**) Arctic and global average, respectively. Red/blue bars are for direct and snow-albedo NRFs, respectively.

4. Conclusions

The present study makes use of a climate-chemistry model with an interactive aerosol module to explore the sensitivity of the Arctic concentration of black carbon aerosols to land transportation emissions from mid-latitude source regions. In particular, the beneficial effects of removing BC emissions from on-road diesel engines are explored by calculating the changes in BC optical depth poleward of 65°N and the radiative impact. A global average decrease of $10.5\% \pm 2\%$ is calculated for BC optical depth at $\lambda = 0.55$ μm, along with a $16.5\% \pm 2\%$ value over the Arctic region, taking into account the seasonal variability. The annual average direct radiative forcing corresponding to these changes is found to be -0.054 W·m^{-2} and -0.074 W·m^{-2}, respectively, which are significant RF values compared to those of other climate agents, in particular the Arctic value. These RF values become significantly larger when considering the snow-albedo effect due to BC wet/dry deposition. According to the model, global and Arctic averages of the annual RF raise up to -0.064 W·m^{-2} and -0.16 W·m^{-2}, respectively.

The rapidly changing regional emission of anthropogenic BC has to be considered as a potential source of uncertainty of the above estimates, as well as possible systematic biases in the emission inventories. If projections for land transportation emissions are used as a robust indication of future trends [61], then there is an indication of a reduction of diesel on-road emissions in 2014 compared to 2000 on the order of 30%. However, the diesel on-road contribution to BC emissions is planned to grow again in the future and return to 2000 values by mid-century, following extensive motorization of highly populated countries in East Asia. A potential future assessment based on multi-model predictions could be useful to narrow the intrinsic uncertainty of the present study, which is related to the model ability to transport pollutants from the source regions (*i.e.*, the Northern Hemisphere mid-latitudes) to the receptor region (*i.e.*, the Arctic). In addition, it should be considered that some non-linear effects on the Arctic mitigation discussed in this work could be expected from year-to-year changes in biomass burning emissions, for which an average climatological value is used here.

Model predictions have first been evaluated using surface BC concentrations observed in three Arctic stations and comparing the calculated high-latitude vertical profile of the BC mass mixing ratio to that observed from aircraft measurements over the Arctic Pacific region. The pronounced seasonality of the calculated Arctic optical thickness is found to be consistent with that from other models (e.g., Koch and Hansen [18]), as well as the calculated differences between eastern and western Arctic values and the role of Eurasian sources in determining a large part of BC loading over the Western Arctic and the North Atlantic. The pronounced seasonal cycle of meridional eddy mass fluxes of BC towards the northern high-latitudes is found to be the most important reason for the BC behavior over the Arctic, in particular for the BC maximum calculated in the model (and observed in the real world) during the Arctic haze period, *i.e.*, late winter and beginning of spring.

On an annual basis, the calculated normalized radiative forcing, *i.e.*, RF normalized to the BC optical depth change at $\lambda = 0.55\ \mu m$, accounts for 110 and 115 over the Arctic and globally, respectively, and increases to 240 and 130 when considering the snow-albedo effect. During springtime months with the highest BC RF, the NRF accounts for 200 and 145 over the Arctic and globally, respectively; these values increase to 460 and 185 when considering the BC/albedo effect. These values are consistent with the annual mean NRF reported in the study of Schulz *et al.* [22] for total atmospheric BC, *i.e.*, 153 ± 68 on a global scale for the direct RF alone, from an average of nine aerosol models. The high BC NRF value over the Arctic region makes large-scale atmospheric transport of BC towards the polar region a significant source of climate forcing; the potential mitigation resulting from the cut-off of on-road diesel emissions should then be considered among those feasible measures to temporarily counteract the effects of growing atmospheric concentrations of long-lived greenhouse gases, CO_2 in particular.

Acknowledgments

Part of this work has been completed under the EC project, REACT4C, Grant No. ACP8-GA-2009-233772. The authors would like to acknowledge David Lee and Ling Lim of Manchester Metropolitan University for providing aircraft emission scenarios to be used in the model runs. The authors would like to thank the anonymous reviewers for their helpful suggestions, mostly one on the correct treatment of the BC single scattering albedo.

Author Contributions

Giovanni Pitari: overall coordination and responsibility for the climate-chemistry-transport modules. Glauco Di Genova: BC emissions and radiative calculations. Natalia De Luca: evaluation of the model results with observations.

Conflicts of Interest

The authors declare no conflict of interest.

References

1. Uherek, E.; Halenka, T.; Borken-Kleefeld, J.; Balkanski, Y.; Berntsen, T.; Borrego, C.; Gauss, M.; Hoor, P.; Juda-Rezler, K.; Lelieveld, J.; *et al.* Transport impacts on atmosphere and climate: Land transport. *Atmos. Env.* **2010**, *44*, 4772–4816.

2. Shine, K.; Forster, P.M. The effect of human activity on radiative forcing of climate change: A review of recent developments. *Glob. Planet. Chang.* **1999**, *20*, 205–225.

3. Intergovernmental Panel for Climate Change (IPCC). Climate change 2013. In *The Physical Science Basis*; Stocker, T.F., Qin, D., Plattner, G.-K., Tignor, M.M.B., Allen, S.K., Boschung, J., Nauels, A., Xia, Y., Bex, V., Midgley, P.M., Eds.; Cambridge University Press: Cambridge, UK, 2013; Chapter 7.

4. Fuller, K.A.; Malm, W.C.; Kreidenweis, S.M. Effects of mixing on extinction by carbonaceous particles. *J. Geophys. Res.* **1999**, *104*, 15941–15954.

5. Saleh, R.; Robinson, E.S.; Tkacik, D.S.; Ahern, A.T.; Liu, S.; Aiken, A.C.; Sullivan, R.C.; Presto, A.A.; Dubey, M.K.; Yokelson, R.J.; *et al.* Brownness of organics in aerosols from biomass burning linked to their black carbon content. *Nat. Geosci.* **2014**, *7*, 647–650.

6. Bond, T.C.; Doherty, S.J.; Fahey, D.W.; Forster, P.M.; Berntsen, T.; DeAngelo, B.J.; Flanner, M.G.; Ghan, S.; Kärcher, B.; Koch, D.; *et al.* Bounding the role of black carbon in the climate system: A scientific assessment. *J. Geophys. Res.* **2013**, *118*, 5380–5552.

7. Cohen, J.B.; Wang, C. Estimating global black carbon emissions using a top-down Kalman filter approach. *J. Geophys. Res.* **2013**, doi:10.1002/2013JD019912.

8. Brandt, E.B.; Kovacic, M.B.; Lee, G.B.; Gibson, A.M.; Acciani, T.H.; Thomas, H.; Le Cras, T.D.; Ryan, P.H.; Budelsky, A.L.; Hershey, G.K.K. Diesel exhaust particle induction of IL-17A contributes to severe asthma. *J. Allergy Clin. Immunol.* **2013**, *132*, 1194–1201.

9. Suglia, S.F.; Gryparis, A.; Wright, R.O.; Schwartz, J.; Wright, R.J. Association of black carbon with cognition among children in a prospective birth cohort study. *Am. J. Epidemiol.* **2008**, *167*, 280–286.

10. McCracken, J.; Baccarelli, A.; Hoxha, M.; Dioni, L.; Melly, S.; Coull, B.; Suh, H.; Vokonas, P.; Schwartz, J. Annual ambient black carbon associated with shorter telomeres in elderly men: Veterans Affairs Normative Aging Study. *Environ. Health Perspect.* **2010**, *118*, 1564–1570.

11. Riddle, S.G.; Robert, M.; Jakober, C.A.; Hannigan, M.P.; Kleeman, M.J. Size-Resolved source apportionment of airborne particle mass in a roadside environment. *Environ. Sci. Technol.* **2008**, *42*, 6580–6586.

12. Twigg, M.W. Progress and future challenges in controlling automotive exhaust gas emissions. *Appl. Catal. B Environ.* **2007**, *70*, 2–15.

13. Bond, T.C.; Streets, D.G.; Yarber, K.F.; Nelson, S.M.; Woo, J.-H.; Klimont, Z. A technology-based global inventory of black and organic carbon emissions from combustion. *J. Geophys. Res.* **2004**, *109*, D14203.

14. Environmental Protection Agency (EPA). *Report to Congress on Black Carbon*; EPA-450/R-12–001; Department of the Interior, Environment, and Related Agencies: Research Triangle Park, NC, USA, 2012.

15. Kopp, R.E.; Mauzeralla, D.L. Assessing the climatic benefits of black carbon mitigation. *PNAS* **2010**, *107*, 11703–11708.

16. World Meteorological Organization (WMO). *Integrated Assessment of Black Carbon and Tropospheric Ozone*; WMO: Geneva, Switzerland, 2011.

17. Pitari, G.; Di Carlo, P.; Coppari, E.; de Luca, N.; Di Genova, G.; Iarlori, M.; Pietropaolo, E.; Rizi, V.; Tuccella, P. Aerosol measurements in central Italy: Impact of local sources and large scale transport resolved by LIDAR. *J. Atmos. Solar-Terr. Phys.* **2013**, *92*, 116–123.

18. Koch, D.; Hansen, J. Distant origins of Arctic black carbon: A Goddard Institute for Space Studies Model E experiment. *J. Geophys. Res.* **2005**, *110*, D04204.

19. Jiao, C.; Flanner, M.G.; Balkanski, Y.; Bauer, S.E.; Bellouin, N.; Berntsen, T.K.; Bian, H.; Carslaw, K.S.; Chin, M.; De Luca, N.; *et al.* An AeroCom assessment of black carbon in Arctic snow and sea ice. *Amos. Chem. Phys.* **2014**, *14*, 2399–2417.

20. Flanner, M.G. Arctic climate sensitivity to local black carbon. *J. Geophys. Res.* **2013**, *118*, 1840–1851.

21. Hansen, J.; Nazarenko, L. Soot climate forcing via snow and ice albedos. *PNAS* **2004**, *101*, 423–428.

22. Schulz, M.C.; Textor, S.; Kinne, Y.; Balkanski, S.; Bauer, T.; Berntsen, T.; Berglen, O.; Boucher, F.; Dentener, S.; Guibert, I.S.A.; *et al.* Radiative forcing by aerosols as derived from the AeroCom present-day and pre-industrial simulations. *Atmos. Chem. Phys.* **2006**, *6*, 5225–5246.

23. Pitari, G.; Mancini, E.; Rizi, V.; Shindell, D.T. Impact of future climate and emission changes on stratospheric aerosols and ozone. *J. Atmos. Sci.* **2002**, *59*, 414–440.

24. Eyring, V.; Butchart, N.; Waugh, D.W.; Akiyoshi, H.; Austin, J.; Bekki, S.; Bodeker, G.E.; Boville, B.A.; Brühl, C.; Chipperfield, M.P.; *et al.* Assessment of temperature, trace species, and ozone in chemistry-climate model simulation of the recent past. *J. Geophys. Res.* **2006**, *111*, D22308.

25. Morgenstern, O.; Giorgetta, M.A.; Shibata, K.; Eyring, V.; Waugh, D.; Shepherd, T.G.; Akiyoshi, H.; Austin, J.; Baumgärtner, A.; Bekki, S.; *et al.* A review of CCMVal-2 models and simulations. *J. Geophys. Res.* **2010**, doi:10.1029/2009JD013728

26. Kärcher, B.; Lohmann, U. A parameterization of cirrus cloud formation: Homogeneous freezing of supercooled aerosols. *J. Geophys. Res.* **2002**, doi:10.1029/2001JD000470.

27. Kärcher, B.; Lohmann, U. A parameterization of cirrus cloud formation: Homogeneous freezing including effects of aerosol size. *J. Geophys. Res.* **2002**, doi:10.1029/2001JD001429.

28. Pitari, G.; Aquila, V.; Kravitz, B.; Robock, A.; Watanabe, S.; Cionni, I.; de Luca, N.; Di Genova, G.; Mancini, E.; Tilmes, S. Stratospheric ozone response to sulfate geoengineering: Results from the Geoengineering Model Intercomparison Project (GeoMIP). *J. Geophys. Res.* **2014**, *119*, 2629–2653.

29. Penner, J.E. Carbonaceous aerosols influencing atmospheric radiation: Black carbon and organic carbon. In *Aerosol Forcing of Climate*; Charlson, R.J., Heintzenberg, J., Eds.; Wiley: Chichester, UK, 1995; pp. 91–108.

30. Koch, D.; Schulz, M.; Kinne, S.; McNaughton, C.; Spackman, J.R.; Bond, T.C.; Balkanski, Y.; Bauer, S.; Berntsen, T.; Boucher, O.; *et al.* Evaluation of black carbon estimations in Global Aerosol Models. *Atmos. Chem. Phys.* **2009**, *9*, 9001–9026.

31. Lamarque, J.F.; Kyle, G.P.; Meinshausen, M.; Riahi, K.; Smith, S.J.; van Vuuren, D.P.; Conley, A.J.; Vitt, F. Global and regional evolution of short-lived radiatively-active gases and aerosols in the representative concentration pathways. *Clim. Chang.* **2011**, *109*, 191–212.

32. Textor, C.; Schulz, M.; Guibert, S.; Kinne, S.; Balkanski, Y.; Bauer, S.; Berntsen, T.; Berglen, T.; Boucher, O.; Chin, M.; *et al.* Analysis and quantification of the diversities of aerosol life cycles within AEROCOM. *Atmos. Chem. Phys.* **2006**, *6*, 1777–1813.

33. Kinne, S.; Schulz, M.; Textor, C.; Guibert, S.; Balkanski, Y.; Bauer, S.E.; Berntsen, T.; Berglen, T.; Boucher, O.; Chin, M.; *et al.* An AeroCom initial assessment—Optical properties in aerosol component modules of global models. *Atmos. Chem. Phys.* **2006**, *6*, 1815–1834.

34. Penner, J.; Hegg, D.; Andreae, M.; Leaitch, D.; Pitari, G.; Annegarn, H.; Murphy, D.; Nganga, J.; Barrie, L.; Feichter, H. Chapter 5: Aerosols and indirect cloud effects. In *Climate Change 2001—IPCC Third Assessment Report*; Houghton, J., Ding, Y., Griggs, D.J., Noguer, M., van der Linden, P.J., Dai, X., Maskell, K., Johnson, C.A., Eds.; Cambridge University Press: Cambridge, UK, 2001; pp. 289–348.

35. Penner, J.E.; Zhang, S.Y.; Chin, M.; Chuang, C.C.; Feichter, J.; Feng, Y.; Geogdzhayev, I.V.; Ginoux, P.; Herzog, M.; Higurashi, A.; *et al.* A comparison of model- and satellite-derived optical depth and reflectivity. *J. Atmos. Sci.* **2001**, *59*, 441–460.

36. Kinne, S.; Lohmann, U.; Feichter, J.; Schulz, M.; Timmreck, C.; Ghan, S.; Easter, R.; Chin, M.; Ginoux, P.; Takemura, T.; *et al.* Monthly averages of aerosol properties: A global comparison among models, satellite data and AERONET ground data. *J. Geophys. Res.* **2003**, doi:10.1029/2001JD001253.

37. Textor, C.; Schulz, M.; Guibert, S.; Kinne, S.; Balkanski, Y.; Bauer, S.; Berntsen, T.; Berglen, T.; Boucher, O.; Chin, M.; *et al.* The effect of harmonized emissions on aerosol properties in global models—An AeroCom experiment. *Atmos. Chem. Phys.* **2007**, *7*, 4489–4501.

38. Koch, D.; Schulz, M.; Kinne, S.; McNaughton, C.; Spackman, J.R.; Balkanski, Y.; Bauer, S.; Berntsen, T.; Bond, T.C.; Boucher, O.; *et al.* Corrigendum to "Evaluation of Black Carbon Estimations in Global Aerosol Models". published in Atmos. Chem. Phys., 9, 9001–9026, 2009. *Atmos. Chem. Phys.* **2010**, *10*, 79–81.

39. Randles, C.A.; Kinne, S.; Myhre, G.; Schulz, M.; Stier, P.; Fischer, J.; Doppler, L.; Highwood, E.; Ryder, C.; Harris, B.; *et al.* Intercomparison of shortwave radiative transfer schemes in global aerosol modeling: Results from the AeroCom Radiative Transfer Experiment. *Atmos. Chem. Phys.* **2013**, *13*, 2347–2379.

40. Chou, M.D.; Suarez, M.J. *A Solar Radiation Parameterization for Atmospheric Studies*; NASA Tech. Rep. TM-1999–104606; NASA Goddard Space Flight Cent.: Greenbelt, MD, USA, 1999.

41. Mishchenko, M.I.; Travis, L.D.; Lacis, A.A. *Scattering, Absorption, and Emission of Light by Small Particles*; Cambridge University Press: Cambridge, UK, 2002.

42. Chipperfield, M.P.; Liang, Q.; Strahan, S.E.; Morgenstern, O.; Dhomse, S.S.; Abraham, N.L.; Archibald, A.T.; Bekki, S.; Braesicke, P.; Di Genova, G.; *et al.* Multi-model estimates of atmospheric lifetimes of long-lived ozone-depleting substances: Present and future. *J. Geophys. Res.* **2014**, *119*, 2555–2573.

43. Schwarz, J.P.; Spackman, J.R.; Gao, R.S.; Watts, L.A.; Stier, P.; Schulz, M.; Davis, S.M.; Wofsy, S.C.; Fahey, D.W. Global-scale black carbon profiles observed in the remote atmosphere and compared to models. *Geophys. Res. Lett.* **2010**, doi:10.1029/2010GL044372.

44. Pueschel, R.F.; Kinne, S.A. Physical and radiative rroperties of Arctic atmospheric aerosols. *Sci. Tot. Env.* **1995**, *160/161*, 811–824.

45. Liousse, C.; Penner, J.E.; Chuang, C.; Walton, J.J.; Eddleman, H.; Cachier, H. A global three-dimensional model study of carbonaceous aerosols. *J. Geophys. Res.* **1996**, *101*, 19411–19432.

46. Quinn, P.K.; Stohl, A.; Arneth, A.; Berntsen, T.; Burkhart, J.F.; Christensen, J.; Flanner, M.; Kupiainen, K.; Lihavainen, H.; Shepherd, M.; *et al. The Impact of Black Carbon on Arctic Climate*; Arctic Monitoring and Assessment Programme (AMAP): Oslo, Norway, 2011.

47. Sharma, S.; Ogren, J.A.; Jefferson, A.; Eleftheriadis, K.; Chan, E.; Quinn, P.K.; Burkhart, J.F. Black Carbon in the Arctic, Arctic Report Card 2013. Available online: http://www.arctic.noaa.gov/reportcard/ (accessed on 12 October 2014).

48. Rosen, H.; Hansen, A.D.A.; Novakov, T. Role of graphitic carbon particles in radiative transfer in the Arctic haze. *Sci. Total Environ.* **1984**, *36*, 103–110.

49. Samset, B.H.; Myhre, G.; Herber, A.; Kondo, Y.; Li, S.-M.; Moteki, N.; Koike, M.; Oshima, N.; Schwarz, J.P.; Balkanski, Y.; *et al.* Modelled black carbon radiative forcing and atmospheric lifetime in AeroCom Phase II constrained by aircraft observations. *Atmos. Chem. Phys.* **2014**, *14*, 12465–12477.

50. Lee, Y.H.; Lamarque, J.-F.; Flanner, M.G.; Jiao, C.; Shindell, D.T.; Berntsen, T.; Bisiaux, M.M.; Cao, J.; Collins, W.J.; Curran, M.; *et al.* Evaluation of preindustrial to present-day black carbon and its albedo forcing from Atmospheric Chemistry and Climate Model Intercomparison Project (ACCMIP). *Atmos. Chem. Phys.* **2013**, *13*, 2607–2634.

51. Cross, E.S.; Onasch, T.B.; Ahern, A.; Wrobel, W.; Slowik, J.G.; Olfert, J.; Lack, D.A.; Massoli, P.; Cappa, C.D.; Schwarz, J.P.; *et al.* Soot particle studies—instrument inter-comparison—project overview. *Aerosol Sci. Technol.* **2010**, *44*, 592–611.

52. Søvde, O.A.; Skowron, A.; Iachetti, D.; Lim, L.; Owen, B.; Hodnebrog, Ø.; Di Genova, G.; Pitari, G.; Lee, D.S.; Myhre, G.; *et al.* Aircraft emission mitigation by changing route altitude: A multi-model estimate of aircraft NO_x emission impact on O_3 photochemistry. *Atmos. Env.* **2014**, doi:10.1016/j.atmosenv.2014.06.049.

53. Hendricks, J.; Kärcher, B.; Lohmann, U.; Ponater, M. Do aircraft black carbon emissions affect cirrus clouds on the global scale? *Geophys. Res. Lett.* **2005**, *32*, L12814.

54. Barrie, L.A. Arctic air pollution: An overview of current knowledge. *Atmos. Environ.* **1986**, *20*, 643–663.

55. Iversen, T.; Joranger, E. Arctic air pollution and large scale atmospheric flows. *Atmos. Environ.* **1985**, *19*, 2099–2108.

56. Breider, T.J.; Mickley, L.J.; Jacob, D.J.; Wang, Q.; Fisher, J.A.; Chang, R.Y.-W.; Alexander, B. Annual distributions and sources of Arctic aerosol components, aerosol optical depth, and aerosol absorption. *J. Geophys. Res.* **2014**, *119*, 4107–4124.

57. Samset, B.H.; Myhre, G.; Schulz, M.; Balkanski, Y.; Bauer, S.; Berntsen, T.K.; Bian, H.; Bellouin, N.; Diehl, T.; Easter, R.C.; *et al.* Black carbon vertical profiles strongly affect its radiative forcing uncertainty. *Atmos. Chem. Phys.* **2013**, *13*, 2423–2434.

58. Hansen, J.; Sato, M.; Ruedy, R.; Nazarenko, L.; Lacis, A.; Schmidt, G.A.; Russell, G.; Aleinov, I.; Bauer, M.; Bauer, S.; *et al.* Efficacy of climate forcings. *J. Geophys. Res.* **2005**, *110*, D18104.

59. Hansen, J.; Sato, M.; Ruedy, R.; Kharecha, P.; Lacis, A.; Miller, R.L.; Nazarenko, L.; Lo, K.; Schmidt, G.A.; Russell, G.; *et al.* Climate simulations for 1880–2003 with GISS ModelE. *Clim. Dyn.* **2007**, *29*, 661–696.

60. Flanner, M.G.; Zender, C.S.; Randerson, J.T.; Rasch, P.J. Present-day climate forcing and response from black carbon in snow. *J. Geophys. Res.* **2007**, *112*, D11202.

61. Walsh, M.P. On and off road policy strategies. In Proceedings of the ICCT International Workshop on Black Carbon, London, UK, 5 January 2009.

8

Summertime Spatial Variations in Atmospheric Particulate Matter and Its Chemical Components in Different Functional Areas of Xiamen, China

Shuhui Zhao [1,2,3], **Liqi Chen** [1,3,*], **Yanli Li** [2], **Zhenyu Xing** [2] **and Ke Du** [4,*]

[1] Key Lab of Global Change and Marine-Atmospheric Chemistry of State Oceanic Administration, Third Institute of Oceanography, State Oceanic Administration, Xiamen 361005, China; E-Mail: shzhao@tio.org.cn

[2] Institute of Urban Environment, Chinese Academy of Sciences, Xiamen 361021, China; E-Mails: liyanli115@126.com (Y.L.); zyxing@iue.ac.cn (Z.X.)

[3] College of Ocean and Earth Sciences and State Key Laboratory of Marine Environmental Science, Xiamen University, Xiamen 361005, China

[4] Department of Mechanical and Manufacturing Engineering, University of Calgary, Calgary, AB T2N 1N4, Canada

* Author to whom correspondence should be addressed; E-Mails: Lqchen@soa.gov.cn (L.C.); kddu@ucalgary.ca (K.D.)

Academic Editors: Huiting Mao and Robert W. Talbot

Abstract: Due to the highly heterogeneous and dynamic nature of urban areas in Chinese cities, air pollution exhibits well-defined spatial variations. Rapid urbanization in China has heightened the importance of understanding and characterizing atmospheric particulate matter (PM) concentrations and their spatiotemporal variations. To investigate the small-scale spatial variations in PM in Xiamen, total suspended particulate (TSP), PM_{10}, PM_5 and $PM_{2.5}$ measurements were collected between August and September in 2012. Their average mass concentrations were 102.50 $\mu g \cdot m^{-3}$, 82.79 $\mu g \cdot m^{-3}$, 55.67 $\mu g \cdot m^{-3}$ and 43.70 $\mu g \cdot m^{-3}$, respectively. Organic carbon (OC) and elemental carbon (EC) in $PM_{2.5}$ were measured using thermal optical transmission. Based on the PM concentrations for all size categories, the following order for the different functional areas studied was identified: hospital > park > commercial area > residential area > industrial area. OC contributed approximately 5%–23% to the $PM_{2.5}$ mass, whereas EC accounted for 0.8%–6.95%. Secondary organic carbon constituted most of the carbonaceous particles found in the park, commercial, industrial and residential areas, with the exception of hospitals. The high PM and EC concentrations in hospitals were primarily caused by vehicle emissions. Thus, the results suggest that long-term plans should be to limit the number of vehicles entering hospital campuses, construct large-capacity underground parking structures, and choose hospital locations far from major roads.

Keywords: particulate matter; organic carbon; elemental carbon; spatial variation; Xiamen; China

---

1. Introduction

Atmospheric particulate matter (PM) is composed of tiny solid particles or liquid droplets that are suspended in the atmosphere. Ambient PM can originate from natural or anthropogenic sources. In urban areas, the ambient PM is primarily of anthropogenic origin and typically comprises a mixture of acids, heavy metals, black carbon, organic chemicals, and soil or dust particles [1]. This pollution can adversely affect the atmospheric environment, human health, and even the global climate. Studies have found that particles can reduce visibility via scattering and absorption, resulting in regional haze pollution events [2,3], which have adverse effects on public health. Epidemiological studies have found that ambient concentrations of PM are associated with various health effects. Both short-term and long-term exposures to airborne PM can result in various respiratory and cardiovascular diseases and even lung cancer. These findings are especially related to fine PM (PM_{10}, *i.e.*, PM with an aerodynamic diameter less than 10 μm, and $PM_{2.5}$, PM with an aerodynamic diameter less than 2.5 μm) [4,5]. Multi-city studies conducted in Europe (29 cities) and in the United States (20 cities) have reported short-term mortality effects for PM_{10} of 0.62% and 0.46% per 10 $\mu g \cdot m^{-3}$ (24-h mean), respectively [6–8]. Raaschou-Nielsen *et al.* [9] found that smaller $PM_{2.5}$ is particularly deadly, identifying a 36% increase in lung cancer per 10 $\mu g \cdot m^{-3}$ increase because these fine particles can penetrate deeper into human lungs.

Environmental protection agencies in many countries have enacted air quality standards to reduce the effects of PM on the environment, human health, and climate. China issued the Ambient Air Quality Standard in 1996 to control air pollution. However, due to the rapid urbanization, population growth and increased vehicle usage, China has experienced serious atmospheric environment problems, particularly PM pollution. Recently, haze resulting from PM pollution has been a particularly acute problem in China [10,11]. Thus, the original ambient air quality standard was revised in 2011 according to the World Health Organization (WHO) air quality guidelines. The standard II (GB 3095-2012) set new limits on PM_{10} and $PM_{2.5}$ concentrations. This regulation limits the 24-h (short-term) and annual (long-term) mean PM_{10} concentrations to 150 $\mu g \cdot m^{-3}$ and 70 $\mu g \cdot m^{-3}$, respectively, while the $PM_{2.5}$ concentrations are limited to 75 $\mu g \cdot m^{-3}$ and 35 $\mu g \cdot m^{-3}$, respectively. Most major cities in China have established state-controlled automatic ambient air quality monitoring stations to monitor air pollution and publish daily urban air quality reports. However, the PM levels observed at one monitoring station are not representative of the entire city [12], especially in Asian countries. Studies have found that the actual exposure levels in Asian countries are much higher than suggested by the measurements collected at monitoring stations due to the special characteristics of Asian residential communities [13,14]. In Asian countries, people have different living styles from those in western countries. For example, people prefer easy access to daily activities. Therefore, residential communities are typically located near the traffic arteries, restaurants, shops, and hospitals. Hence, people spend more time near the PM sources, resulting in a high variability in pollutant levels within an urban area. Taking China as an example, because its urban areas are typically compact, each functional area is not large, although they do have dense populations. Due to the highly density of anthropogenic activities, such as widespread traffic, cooking, and other emission sources, e.g., restaurants and temples, air quality is highly variable between different functional areas. Thus, to more accurately estimate PM exposure, identify potential sources, and effectively protect public health by proposing targeted control strategies, the fine-scale spatial variations in PM pollution must be better understood.

In this study, small-scale spatial variations in PM concentrations were observed in Xiamen between August and September in 2012. Xiamen is one of the cleanest coastal cities in China. However, due to rapid urbanization, the air quality of Xiamen has gradually deteriorated. Research data has shown that hazy days increased by approximately six-fold from 2003 to 2008 [15]. Most of the studies in Xiamen have focused on long-term PM pollution in one or two specific locations [16–20]; no studies have addressed small-scale spatial variations of short-term PM exposure across the different functional urban areas. Thus, the objective of this study is to assess the short-term PM levels, characterize the chemical components of $PM_{2.5}$, and identify the potential sources and spatial variations of PM between different functional areas in Xiamen. The results will provide a scientific foundation for assessing exposure levels and knowledge-based urban planning to protect public health.

2. Materials and Methods

2.1. Site Description

Xiamen, which is a coastal city located in southeastern China, is a modern international portal city for tourism and is one of the earliest participants in China's opening-up policy as a special economic zone (Figure 1). Xiamen comprises Xiamen Island, Gulangyu Island, and part of the rugged mainland coastal region from the left bank of the Jiulong River in the west to the islands of Xiang'an in the northeast. Lying in the temperate and subtropical zone, Xiamen has a subtropical oceanic monsoon climate that is indicative of a mild climate, abundant rainfall, and long summers which typically last from May to October. The annual mean temperature is approximately 21 °C, the annual precipitation is approximately 1200 mm, and the prevailing wind is from the northeast. The city's urban area includes the old urban island area and covers all six districts of Xiamen (i.e., Huli, Siming, Jimei, Tong'an, Haicang and recently Xiang'an), with a total urban population of 1,861,289. Xiamen Island is divided into two parts, the Siming and Huli districts. The Siming and Huli districts form the special economic zone. Although the area of Xiamen Island is only 7.5% of the total area of Xiamen City, it has an urban population that exceeds 884,100, and accounts for approximately 47.7% of the total population. Thus, our study focused on the urban area on Xiamen Island.

In this work, PM samples were collected from the ambient atmosphere of five different functional areas in Xiamen Island. The five different functional areas included parks, hospitals, commercial areas, industrial areas and residential areas, which are very relevant to our daily lives. To capture sufficient variability, 16 sites were selected from the Xiamen Island urban area in China. For each functional area, three or four monitoring sites were selected to collect aerosol samples to study PM characteristics. These sampling sites were selected according to their land-use types. All of the sampling sites are shown in Figure 1, and the specific latitude and longitude for each site are listed in Table 1.

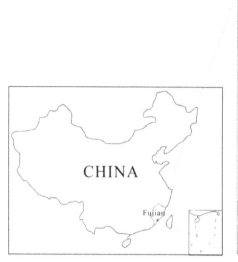
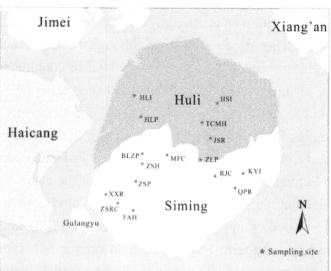

Figure 1. Map of Xiamen City and sampling sites (**Parks**: Zhongshan Park (ZSP), Bailuzhou Park (BLZP), Zhonglun Park (ZLP), and Huli Park (HLP); **Hospitals**: First Affiliated Hospital of Xiamen University (FAH), Zhongshan Hospital of Xiamen University (ZSH), and Xiamen Hospital of Traditional Chinese Medicine (TCMH); **Commercial areas**: Zhongshan Road commercial area (ZSRC), Ruijing commercial area (RJC), and Mingfa commercial area (MFC); **Industrial areas**: Kaiyuan industrial area (KYI), Huli industrial area (HLI), and Heshan industrial area (HSI); and **Residential areas**: Xiaxi residential area (XXR), Qianpunan residential area (QPR), and Jinshang residential area (JSR)).

Table 1. Description of the sampling sites and meteorological conditions.

Sampling Sites		Locations	Sampling Dates	Meteorological Parameters			
				W.S. (m·s⁻¹)	T (°C)	R.H. (%)	P (hPa)
Parks	ZSP	24°27′40.82″N 118°05′04.17″E	8–7	0.3	32.0	66.4	995.6
	BLZP	24°28′50″N 118°05′35″E	8–14	0.4	31.0	68.3	1002.7
	ZLP	24°29′05″N 118°08′35″E	8–20	0.3	28.4	83.4	1001.8
	HLP	24°30′24.65″N 118°06′05.74″E	8–30	0.2	29.5	77.5	1017.1
Hospitals	FAH	24°27′16″N 118°05′00″E	8–8	0.1	31.6	67.1	994.3
	ZSH	24°28′27.34″N 118°05′34.71″E	8–13	0.3	30.2	73.0	1003.0
	TCMH	24°30′11.39″N 118°08′09.36″E	8–29	0.6	0.5	0.0	28.9

Table 1. *Cont.*

Sampling Sites		Locations	Sampling Dates	Meteorological Parameters			
				W.S. (m·s⁻¹)	T (°C)	R.H. (%)	P (hPa)
Commercial areas	ZSRC	24°27′24″N 118°04′40″E	8–10	0.6	31.0	71.8	1000.6
	RJC	24°28′40.78″N 118°09′16.85″E	8–21	0.2	31.6	64.3	1013.6
	MFC	24°28′55″N 118°07′2.8″E	9–4	0.6	31.8	62.6	1023.0
Industrial areas	KYI	24°28′50″N 118°10′17″E	8–22	0.8	31.9	59.8	1012.8
	HLI	24°30′54″N 118°05′39″E	8–28	0.7	30.6	68.9	1014.2
	HSI	24°30′53″N 118°08′55″E	9–6	0.9	30.8	67.2	1021.4
Residential areas	XXR	24°27′33″N 118°04′30″E	9–5	0.6	31.8	62.6	1023.0
	QPR	24°28′8.4″N 118°10′43″E	9–11	0.2	30.3	72.8	1018.3
	JSR	24°29′44.45″N 118°08′45.66″E	9–12	0.5	30.3	72.5	1015.5

Notes: **Parks**: ZSP = Zhongshan Park, BLZP = Bailuzhou Park, ZLP = Zhonglun Park, and HLP = Huli Park; **Hospitals**: FAH = First Affiliated Hospital of Xiamen University, ZSH = Zhongshan Hospital of Xiamen University, and TCMH = Xiamen Hospital of Traditional Chinese Medicine; **Commercial areas**: ZSRC = Zhongshan Road commercial area, RJC = Ruijing commercial area, and MFC = Mingfa commercial area; **Industrial areas**: KYI = Kaiyuan industrial area, HLI = Huli industrial area, and HIS = Heshan industrial area; **Residential areas**: Xiaxi residential area (XXR), QPR = Qianpunan residential area, and JSR = Jinshang residential area.

2.2. Sample Collection

A total of 61 PM samples, including 15 TSP, 15 PM_{10}, 15 PM_5, and 16 $PM_{2.5}$ samples, were collected from August to September in 2012 (the PM sample collected from Zhonglun Park was only a $PM_{2.5}$ sample). The collection sites were selected not only according to their functional characteristics but also based on certain practical limitations for locating the sampling device (*i.e.*, the topography of the ground and the availability of electrical power outlets). The sampling locations were 3 m to 12 m above ground level, where most human activities occur. In this study, PM samples were collected on quartz fiber filters (90-mm diameter, Whatman®) at a flow rate of 100 L/min for 24 h using mid-volume samplers (TH1500, Wuhan Tianhong Instrument Co., Ltd, Wuhan, China). To remove residual carbon, the quartz filters were pre-combusted at 600 °C for 4 h in a furnace before sampling. After sampling, the filters were removed from the filter holder and wrapped with baked aluminum foil, sealed within polyethylene plastic bags, and stored at 4 °C before analysis. When loading or unloading the filters, the operator wore plastic gloves to minimize contamination. No samples were collected during rain events.

An automatic weather station (Kestrel 4500, USA) was co-located with the sampler. The weather station recorded daily and hourly measurements of air temperature, wind speed and direction, air pressure, and relative humidity with measurement resolutions (including uncertainties) of 0.1 °C (±1 °C), 0.1 m/s (±3%), 1° (±5°), 0.1 hPa (±1.5 hPa), and 0.1% R.H. (±3%), respectively. The observed meteorological parameters are summarized in Table 1. Figure 2 presents the wind direction and wind speed frequencies during the sampling period. Meteorological data, such as precipitation and visibility, were obtained from the weather website Weather Underground [21]. The Air Pollution Index (API), which is a parameter used to describe air quality, was downloaded from the Ministry of Environmental Protection of the People's Republic of China. The daily variations in these parameters during the sampling period are shown in Figure 3.

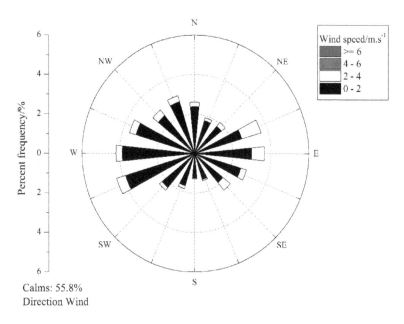

Figure 2. Wind direction and wind speed frequencies during the sampling period.

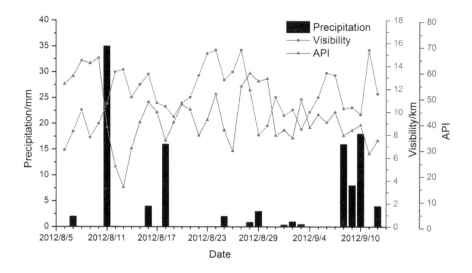

Figure 3. Daily variations in visibility, API and precipitation during the sampling period in Xiamen.

2.3. Sample Analyses

Quartz fiber filters were weighed with a Sartorius microbalance before and after sampling to obtain the masses of the atmospheric particulate matter. Before weighing, all filters were equilibrated for 24 h under a controlled temperature and relative humidity conditions (25 °C and 50%) in an electronic humidity chamber. The PM mass concentrations were calculated by dividing the mass by the sampling volume.

2.4. OC and EC Analysis

To determine the OC and EC concentrations in the $PM_{2.5}$ samples, a 1.6-cm^2 patch of filter was cut from each $PM_{2.5}$ sample and analyzed with an automated semi-continuous thermal-optical transmittance (TOT) carbon analyzer (Sunset Laboratory, Model-4, USA). The TOT method follows the National Institute for Occupational Safety and Health (NIOSH) protocol [22]. The following provides a brief description of the method. First, the small piece of the filter is gradually heated in a pure helium (He) gas environment to 850 °C, at which point most organic carbon is thought to be converted into carbon dioxide (CO_2), whereas small organic compounds are charred into EC. The CO_2 is then swept out of the oven by He gas and detected by a self-contained non-dispersive infrared (NDIR) system. Then, after the sample oven has cooled to 550 °C, the sample is step-heated to 870 °C in an oxidizing environment of 2% oxygen-containing helium (2% O_2, 98% He). At this stage, EC is oxidized into CO_2 and detected by the NDIR. Finally, a standard methane gas is injected for reference. The split between OC and EC is corrected when the laser transmittance returns to the initial value. The precision of the instrument is ± 5%, and the detection limit is 0.1 µg·cm^{-2}.

2.5. Analysis of Water-Soluble Inorganic Ions

Another 1.6-cm^2 patch of filter was cut from each $PM_{2.5}$ sample to analyze the water-soluble inorganic ions. To sufficiently extract the water-soluble ions from the filters, each filter was extracted by 10 mL aliquots of ultra-deionized water for 30 min using an ultrasonic water bath device. The extracts were filtered through 0.45-µm micro-porous filters and stored in 2 mL sample vials. Subsequently, three major anions (Cl^-, NO_3^- and SO_4^{2-}) of the $PM_{2.5}$ samples were analyzed using a Dionex Ion Chromatograph (ICS3000, Dionex, USA). The chromatography conditions for anions were as follows: IonpacAS23 Analytical column: 4 × 250 mm, anion self-regretting suppressor ASRS 300: 4 mm, Auto suppression recycle mode, eluent: 4.5 mmol/L Na_2CO_3 and 0.8 mmol/L $NaHCO_3$ flow rate: 1.0mL/min, and volume of injected sample: 20 µL.

3. Results and Discussion

3.1. Mass Concentrations of Atmospheric Particles

Within each functional area, three or four locations were selected to sample the ambient aerosols to study their PM characteristics. To provide a general description of the PM levels, the TSP, PM_{10}, PM_5 and $PM_{2.5}$ mass concentrations are listed in Table 2. The PM_{10}/TSP and $PM_{2.5}$/PM_{10} ratios are also listed in Table 2. The TSP concentrations in Xiamen were found to range from 63.51 µg·m^{-3} to 142.02 µg·m^{-3}, with a mean of 102.50 ± 23.14 µg·m^{-3}. The PM_{10} concentrations varied from 51.41 µg·m^{-3} to 116.95 µg·m^{-3}, with a mean of 82.79 ± 20.79 $\mu g \cdot m^{-3}$. The $PM_{2.5}$ concentrations ranged from 22.49 to 70.38 $\mu g \cdot m^{-3}$, with a mean of 43.70 ± 15.23 $\mu g \cdot m^{-3}$. The mean PM_{10}/TSP ratio was approximately 0.806, which indicates that most of the TSP particles were inhalable particles smaller than 10 μm. The average $PM_{2.5}/PM_{10}$ ratio was 0.528, which is higher than that stipulated by the Chinese Ambient Air Quality Standards (GB 3095-2012; in the new standard II, $PM_{2.5}$ should account for no more than 50% of the ambient PM_{10}), especially near the hospitals and parks, which were near parking lots. These observations suggest that $PM_{2.5}$ in Xiamen may come from traffic vehicles. Xiamen is among the top 10 cleanest cities in China. All of the PM_{10} and $PM_{2.5}$ mass concentrations were below the 24-hr mass-based standards of 150 $\mu g \cdot m^{-3}$ and 75 $\mu g \cdot m^{-3}$ set by the Chinese Ambient Air Quality Standards (GB 3095-2012, the new standard II). However, the concentrations exceeded the 24-hr mass-based standards of 50 $\mu g \cdot m^{-3}$ (PM_{10}) and 25 $\mu g \cdot m^{-3}$ ($PM_{2.5}$) set by the World Health Organization's Healthy Ambient Air quality standards. The higher ambient average concentrations of small particulates (PM_{10} and $PM_{2.5}$) suggest that inhalable particles, which have been linked to comparatively higher rates of premature mortality in many epidemiological studies, were the primary pollutants affecting human health during the study period in Xiamen.

Table 2. Mass concentrations measured from the PM samples collected at different functional areas in Xiamen.

Sampling Sites		Mass Concentrations ($\mu g \cdot m^{-3}$)				Ratios (%)	
		TSP	PM_{10}	PM_5	$PM_{2.5}$	PM_{10}/TSP	$PM_{2.5}/PM_{10}$
Parks	ZSP	123.16	106.98	85.50	66.08	86.86	61.77
	BLZP	79.01	60.79	40.45	27.64	76.94	45.48
	ZLP	-	-	-	52.05	-	-
	HLP	119.19	100.70	76.12	47.87	84.49	47.54
	Means	**107.12**	**89.49**	**67.36**	**48.41**	**82.76**	**51.60**
Hospitals	FAH	142.02	116.95	91.64	70.38	82.35	60.18
	ZSH	90.14	72.13	50.16	34.60	80.02	47.97
	TCMH	120.48	109.61	86.41	68.83	90.98	62.80
	Means	**117.55**	**99.56**	**76.07**	**57.94**	**84.45**	**56.98**
Commercial areas	ZSRC	126.72	96.61	69.17	46.84	76.24	48.48
	RJC	91.31	70.47	49.93	34.30	77.17	48.68
	MFC	101.12	84.43	60.37	41.78	83.49	49.48
	Means	**106.38**	**83.84**	**59.82**	**40.97**	**78.97**	**48.88**
Industrial areas	KYI	-	66.57	48.49	33.83	-	50.82
	HLI	63.51	51.41	35.82	22.49	80.95	43.74
	HSI	97.52	77.65	53.08	31.66	79.62	40.77
	Means	**80.52**	**65.21**	**45.80**	**29.33**	**80.29**	**45.11**
Residential areas	XXR	106.26	84.19	60.42	39.20	79.23	46.56
	QPR	65.70	51.73	36.98	27.10	78.73	52.39
	JSR	108.90	91.60	69.60	54.50	84.11	59.50
	Means	**93.62**	**75.84**	**55.67**	**40.27**	**80.69**	**52.82**
Means		102.50	82.79	60.94	43.70	81.51	51.07
SDs		23.14	20.79	18.22	15.23	4.16	6.84

The average PM mass concentrations for the five functional areas exhibited the following order: hospitals > parks > commercial areas > residential areas > industrial areas. Among the 16 sites, the lowest observed PM levels in Xiamen were found in the Huli industrial area, whereas the highest PM values were observed at the First Affiliated Hospital of Xiamen University (FAH). At FAH, $PM_{2.5}$ accounted for approximately 60% of the PM_{10}, which nearly reached 75 $\mu g \cdot m^{-3}$. The $PM_{2.5}$ concentrations in Zhongshan Park, Xiamen Hospital of Traditional Chinese Medicine (TCMH) and Jinshang residential area were also higher than in the other areas. According to the study by the authors [23], emissions from traffic vehicles are an important source of $PM_{2.5}$. The selected sampling sites were as far from the pollution sources as possible as to ensure that the sites were representative of the corresponding functional areas. However, traffic vehicles cannot be avoided in urban areas, especially near hospitals, parks, and commercial and residential areas, which are closely related to people's daily lives. Thus, one reason for the higher PM pollution in these areas is the emissions. There are no heavy industries on Xiamen Island. The existing light industries, such as the optoelectronics and software industries, emit small amounts of PM into the atmosphere. Meanwhile, these industrial areas are relatively large with relatively few vehicle emissions and other pollution sources. Thus, the lowest concentrations were found in the industrial areas.

To identify the impact on the ambient air quality, correlation coefficients between the PM, API and visibility were calculated; the results are listed in Table 3. Because the API was calculated based on the concentrations of five atmospheric pollutants, i.e., PM_{10}, sulfur dioxide (SO_2), nitrogen dioxide (NO_2), carbon monoxide (CO), and ozone (O_3), measured at the monitoring stations throughout the city, the API exhibited significant positive correlation coefficients with PM_5 (r = 0.599 at p = 0.05), PM_{10} (r = 0.588 at p = 0.05), and TSP (r = 0.578 at p = 0.05), although the same correlation was not identified for $PM_{2.5}$. These observations indicate that the mass concentrations of PM collected at different areas in our study were similar to the results of the environmental protection departments. Because high visibility is common when the air quality is good, significant negative correlation coefficients were identified between the API and visibility (r = −0.441 at p = 0.01). Significant negative correlation coefficients were also found between visibility and PM_5 (r = −0.772 at p = 0.01), PM_{10} (r = −0.778 at p = 0.01), and TSP (r = −0.728 at p = 0.01), although this was not true for $PM_{2.5}$. These relationships suggest that visibility was primarily reduced via the scattering light due to the presence of larger particles (particulate matter with an aerodynamic diameter greater than or equal to 5.0 μm), such as PM_5 and PM_{10}, during the summertime observation period in Xiamen. Non-significant correlations between $PM_{2.5}$ and both the API and visibility suggest that $PM_{2.5}$ did not play a primary role in visibility and the API during the investigated period. Table 2 and Figure 3 show that the air quality was good during this period.

In this study, samples were not collected simultaneously; the meteorological conditions may be slightly different. Viability in the meteorological parameters was tested using the one-way analysis of variance (ANOVA) method. Significant (p < 0.05) differences in wind speed were found between the five functional areas (F = 3.815 at p = 0.035) during the sampling periods, whereas no significant differences for temperature (F = 0.745 at p = 0.581), relative humidity (F = 1.27 at p = 0.339), and pressure (F = 2.138 at p = 0.144) were found during the sampling periods between the five functional areas. The wind speed is usually considered to have a very significant effect on the diffusion of particles. Thus, to study the influence of the meteorological conditions on PM, we also calculated the correlations between PM and wind speed, relative humidity and temperature; the results are shown in Table 3.

However, non-significant correlation coefficients were found between wind speed and PM levels in this study. Figure 2 shows that most of the wind speeds were less than 2.0 m·s^{-1} (approximately 55.8% of the measurements were calm) at different sampling sites during the sampling periods. This finding suggests that the low wind speeds resulted in weak diffusion of PM$_{2.5}$; therefore, the wind had little effect on the mass concentrations of PM. Precipitation usually exerts a very significant scavenging effect on the airborne particles. Figure 3 shows that there were several precipitation events between 5 August and 12 September 2012. Although atmospheric particles were not collected on rainy days, PM (and pollutant concentrations) may be affected by the previous day's rain. Therefore, the correlation between precipitation and PM level was also calculated. Non-significant correlation coefficients were also found between precipitation events and PM levels in this study. These findings indicate that the variations in meteorological conditions had little effect on the PM levels at the different sites, although the meteorological conditions exhibited small variations during the study period. Thus, we compared our results primarily according to the different functional areas.

Table 3. Relationships between PM, air quality and meteorological parameters during the sampling period.

	Precip	R.H	W.S.	T	Vis	API	PM$_{2.5}$	PM$_5$	PM$_{10}$	TSP
Precip	1	0.199	0.155	−0.188	−0.162	−0.155	0.282	0.444	0.399	0.277
R.H		1	−0.328	0.111	−0.220	0.083	−0.226	0.251	0.249	0.081
W.S.			1	−0.927 **	0.138	−0.024	−0.049	−0.254	−0.221	−0.157
T				1	0.128	0.039	0.338	−0.088	−0.086	0.100
Vis					1	−0.441 **	−0.146	−0.772 **	−0.778 **	−0.728 **
API						1	−0.105	0.599 *	0.588 *	0.578 *
PM$_{2.5}$							1	−0.050	−0.052	−0.016
PM$_5$								1	0.992 **	0.953 **
PM$_{10}$									1	0.979 **
TSP										1

Notes: Precip = Precipitation, Vis = Visibility, R.H. = Relative humidity, W.S. = Wind speed, T = Temperature, and API = Air Pollution Index; * $p = 0.05$ and ** $p = 0.01$.

3.2. Chemical Components in PM$_{2.5}$

According to several studies [9,24], most health problems are caused by the exposure to PM$_{2.5}$ rather than PM$_{10}$. Thus, chemical components of this pollutant should be carefully analyzed and studied. In this study, the chemical components of PM$_{2.5}$, such as organic carbon (OC) and elemental carbon (EC), were analyzed via thermal-optical transmission. To investigate the relationship between the carbonaceous components, sulfate and nitrate, the water-soluble inorganic anions in PM$_{2.5}$ were also determined. The distinct characteristics are described as follows.

3.2.1. Concentrations of OC, EC and Anions in PM$_{2.5}$

Carbonaceous aerosols, which are composed of OC and EC, are an important component of atmospheric particulate matter, especially PM$_{2.5}$. The mass concentrations of OC and EC in PM$_{2.5}$ are listed in Table 4 based on our observations. The average OC and EC concentrations were 4.79 ± 1.19 µg·m^{-3} and

1.09 ± 0.55 µg·m^{-3}, respectively, during the sampling period. The relative standard deviations (RSD = 100% × standard deviation/average) of OC and EC were 25% and 51%, respectively, which are similar to the values reported by Cao et al. [25] for OC (36%) and EC (43%) during summer in Xiamen. The OC concentrations varied from 2.53 to 6.62 µg·m^{-3}, representing approximately 5%–23% of PM$_{2.5}$. The EC concentrations ranged from 0.26 to 2.44 µg·m^{-3}, accounting for 0.8%–6.95% of the total mass of PM$_{2.5}$ mass. Combined, OC and EC contributed approximately 7%-30% of the PM$_{2.5}$ mass, which differs slightly from the pollution characteristics of the Pearl Delta River Region of China in summer [26]. According to Turpin and Lim [27], the ratio of organic matter (OM) to organic carbon (OC) is approximately 1.6 for urban aerosols. Thus, the OM concentrations were calculated as OM = 1.6 × OC, and carbonaceous matter (CM) was estimated as the sum of OM and EC [20,25]. As shown in Table 4, the average OM and CM concentrations in PM$_{2.5}$ were 7.66 ± 1.90 µg·m^{-3} and 8.75 ± 2.45 µg·m^{-3}, respectively, thereby representing average contributions of 19.04% and 21.69% of PM$_{2.5}$, respectively. The organic pollutant levels in PM$_{2.5}$ are lower than those in Guangzhou [26].

Table 4. Mass concentrations of OC, EC, and anions during the sampling period at different sites in Xiamen.

Sampling Sites		Mass Concentrations (µg·m^{-3})							Ratios (%)			
		OC	EC	Cl$^-$	SO$_4^{2-}$	NO$_3^-$	OM	CM	OC/PM$_{2.5}$	EC/PM$_{2.5}$	OM/PM$_{2.5}$	CM/PM$_{2.5}$
Parks	ZSP	6.62	1.2	0.16	9.60	2.21	10.6	11.8	10.02	1.82	16.03	17.85
	BLZP	3.52	1.12	0.31	3.61	2.33	5.63	6.75	12.72	4.04	20.36	24.41
	ZLP	4.16	0.43	0.16	5.53	2.55	6.65	7.09	7.99	0.83	12.78	13.61
	HLP	2.53	1.17	0.6	6.63	2.44	4.06	5.23	5.3	2.45	8.47	10.92
	Means	**4.21**	**0.98**	**0.31**	**6.34**	**2.38**	**6.73**	**7.71**	**9.01**	**2.29**	**14.41**	**16.7**
Hospitals	FAH	6.54	2.44	0.17	12.10	2.33	10.46	12.91	9.29	3.47	14.87	18.34
	ZSH	3.08	1.18	0.04	1.20	3.55	4.93	6.11	8.91	3.4	14.25	17.65
	TCMH	4.97	1.95	0.22	11.50	2.65	7.96	9.9	7.22	2.83	11.56	14.39
	Means	**4.86**	**1.86**	**0.14**	**8.27**	**2.84**	**7.78**	**9.64**	**8.47**	**3.23**	**13.56**	**16.79**
Commercial areas	ZSRC	6.1	1.12	0.27	4.85	2.64	9.76	10.89	13.03	2.4	20.85	23.25
	RJC	4.58	0.47	0.16	3.49	2.19	7.33	7.8	13.36	1.38	21.37	22.75
	MFC	5.47	0.97	0.13	4.60	2.54	8.76	9.73	13.1	2.32	20.96	23.28
	Means	**5.39**	**0.86**	**0.19**	**4.31**	**2.46**	**8.62**	**9.48**	**13.16**	**2.03**	**21.06**	**23.09**
Industrial areas	KYI	5.36	0.8	0.23	1.66	2	8.58	9.38	15.85	2.37	25.35	27.72
	HLI	5.19	1.56	0.13	6.95	2.33	8.3	9.86	23.07	6.95	36.91	43.86
	HSI	5.21	1.09	0.04	1.78	1.91	8.33	9.42	16.45	3.44	26.31	29.75
	Means	**5.25**	**1.15**	**0.13**	**3.46**	**2.08**	**8.4**	**9.55**	**18.45**	**4.26**	**29.52**	**33.78**
Residential areas	XXR	4.3	0.93	0.08	3.10	1.88	6.88	7.81	10.97	2.38	17.55	19.93
	QPR	3.66	0.26	0.24	2.23	2.12	5.85	6.11	13.49	0.96	21.59	22.55
	JSR	5.27	0.72	0.17	6.51	2.45	8.44	9.16	9.68	1.32	15.48	16.80
	Means	**4.41**	**0.64**	**0.17**	**3.95**	**2.15**	**7.06**	**7.70**	**11.38**	**1.55**	**18.21**	**19.76**
Means		4.79	1.09	0.19	5.33	2.38	7.66	8.75	11.9	2.65	19.04	21.69
SDs		1.19	0.55	0.13	3.39	0.39	1.9	2.45	4.28	1.48	6.85	7.79

The highest and lowest observed EC values in Xiamen were found at the First Affiliated Hospital and Qianpunan residential area, respectively. The highest and lowest OC values were found in Zhongshan

Park and Huli Park, respectively. The average EC mass concentrations in the five functional areas exhibited the following order: hospital > industrial area > park > commercial area > residential area. In contrast, OC and OM had the following order: commercial area > industrial area > hospital > residential area > park. Moreover, CM exhibited the following order: hospital > industrial area > commercial area > park > residential area. The average $EC/PM_{2.5}$ ratios had the following order: industrial area > hospital > park > commercial area > residential area, whereas the $CM/PM_{2.5}$ ratios were found to exhibit the following order: industrial area > commercial area > residential area > hospital > park. Based on the above analysis, it can be found that EC concentrations are lower in residential areas. As a developed city, the fuel structure for cooking has switched to natural gases and central stream, thus, EC form cooking sources contributed little at the residential areas in Xiamen. The $OC/PM_{2.5}$ and $OM/PM_{2.5}$ ratios had the following order: industrial area > commercial area > residential area > park > hospital, which differed from OC, EC and $PM_{2.5}$. This result suggests carbonaceous aerosol pollutions cannot be ignored, although the PM pollution in the industrial area was not obviously due to the light industry in Xiamen.

Major inorganic anions (e.g., Cl^-, SO_4^{2-}, and NO_3^-) in $PM_{2.5}$ are also shown in Table 4. Of all the identified anions, sulfate was the dominant component, followed by nitrate; the concentration of chloride was much lower. The concentrations of SO_4^{2-} varied from 1.20 to 12.1 $\mu g \cdot m^{-3}$, with a mean value of 5.33 ± 3.39 $\mu g \cdot m^{-3}$, and the concentration of NO_3^- varied from 1.88 to 3.55 $\mu g \cdot m^{-3}$, with a mean value of 2.38 ± 0.39 $\mu g \cdot m^{-3}$. The mean concentration of Cl^- was approximately 0.19 ± 0.13 $\mu g \cdot m^{-3}$.

3.2.2. Spatial Variations in PM and its Chemical Components

To evaluate spatial similarities and differences, the mean $PM_{2.5}$, OC, EC, anions, OM and CM concentrations for the five functional areas were calculated using a multivariate general linear model (GLM). A GLM is a generalization of multiple linear regression models to accommodate more than one dependent variable. The results are listed in Table 5, in which the F statistic values and significance values are also reported. We specified a significance value ($p < 0.05$) as the threshold for statistically significant differences between the dependent variables, which is used in other statistic methods, such as Student's T-tests [28,29]. Contini et al. [29] found that $PM_{2.5}$ exhibited significant differences between different types of sites in Italy. $PM_{2.5}$ levels increased from background sites to industrial sites and to urban sites, and the $PM_{2.5}/PM_{10}$ ratio was significantly lower (0.61) at background sites compared to industrial and urban sites. Although the $PM_{2.5}/PM_{10}$ ratio in Xiamen exhibited the following order: hospital > residential area > park > commercial area > industrial area, $PM_{2.5}$ exhibited no significant differences ($p = 0.200$) between the five functional areas. These results demonstrate that the EC concentrations in the five functional areas exhibited significant differences during the study period ($p = 0.04$), whereas OC, Cl^-, SO_4^{2-}, NO_3^-, OM and CM had no significant differences. These results are similar to studies in European cities. Putaud et al. [30] based on numerous measurements, found that EC, SO_4^{2-} and NO_3^- exhibited different characteristics for different site types in Italy. For example, they found that the contribution of EC to PM_{10} increased from rural sites to urban and curbside areas, whereas OM was remarkably similar for sites (at least when present). Sandrini et al. suggested that the more marked spatial variability for EC rather than for OC and other components is due to the primary only nature of EC emissions [28]. Table 5 shows that the $OC/PM_{2.5}$ ratios exhibited significant differences ($p = 0.003$), although there were no significant differences between the OC concentrations in the five areas. One possible explanation for this discrepancy is that EC and OC may be correlated with the local emissions in the different functional areas, whereas other components may be from other sources with few local emissions. Additionally, because the database is limited in terms of its temporal span and number of samples, the conclusions may be influenced by these limitations.

Table 5. Results of the GLM applied to $PM_{2.5}$ and its chemical components (tests of between-subjects effects) for the five functional areas in Xiamen.

Source	Dependent Variable	Type III Sum of Squares	df	Mean Square	F	Sig
	$PM_{2.5}$	1374.230	4	343.557	1.796	0.200
	OC	3.506	4	0.877	0.548	0.705
	EC	2.607	4	0.652	**3.632**	**0.040**
	Cl^-	0.073	4	0.018	1.071	0.416
Functional	NO_3^-	1.090	4	0.273	2.423	0.111
Area	SO_4^{2-}	49.267	4	12.317	1.098	0.405
	OM	8.963	4	2.241	0.547	0.705
	CM	13.496	4	3.374	0.650	0.639
	$OC/PM_{2.5}$	203.261	4	50.815	**7.797**	**0.003**
	$EC/PM_{2.5}$	14.015	4	3.504	2.039	0.158

Notes: $p = 0.05$; df = Degree of freedom; F = F statistic value; Sig = Significance (p value). The values in bold are statistically significant differences ($p < 0.05$).

To identify and separate the impact of various sources, correlation coefficients between OC, EC and individual ionic species were calculated and are shown in Table 6. Significant positive correlation coefficients were found between SO_4^{2-} and EC ($r = 0.706$ at $p = 0.01$) which suggests that SO_4^{2-} and EC may share the same sources. EC particles are emitted from incomplete combustion of biofuels, fossil fuels and biomass and are typically considered to be the primary pollutant emitted by the traffic vehicles in urban areas. The oxidation of SO_2 emitted from coal burning for industries is an important source of SO_4^{2-}. Heavy fuel oil, which has high sulfur consent, is commonly used for ship combustion, which may contribute to SO_2 emissions and may represent a possible source of SO_4^{2-}. Xiamen is a low industrial coal consumption city, but the increased shipping emissions from harbor areas are very intense during summertime. Therefore, in addition to industrial emissions, these findings suggest that a portion of EC and SO_4^{2-} in Xiamen may have also been emitted by ship transportation during the study period.

Table 6. Inter-species correlation coefficients.

	Cl^-	SO_4^{2-}	NO_3^-	OC	EC
Cl^-	1	0.190	−0.041	−0.38	0.009
SO_4^{2-}		1	0.053	0.471	0.711 **
NO_3^-			1	−0.271	0.185
OC				1	0.377
EC					1

Notes: * $p = 0.05$ and ** $p = 0.01$.

The relationship between OC and EC provides some indication of the origins of carbonaceous PM$_{2.5}$ [25]. The non-significant correlation coefficients between OC and EC (r = 0.377 at p > 0.05) indicate that the sources of OC and EC are likely to be different and complex. Figure 4 shows the relationship between EC and OC in PM$_{2.5}$ in the different functional areas of Xiamen during the summer study period. EC and OC were well correlated in the samples collected from the hospitals (R^2 = 0.99) and commercial areas (R^2 = 0.92), which suggests a combination of common source contributions. Poor correlations were found for samples from the parks, industrial areas and residential areas, indicating more complex emission sources over these areas.

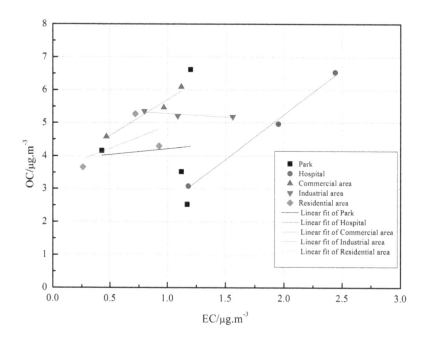

Figure 4. Correlation between EC and OC in PM$_{2.5}$ for the different functional areas of Xiamen during the summer study period; the Adj. R^2 of Parks, Hospitals, Commercial areas, Industrial areas and Residential areas were −0.49, 0.99, 0.92, 0.45, −0.32 and −0.49, respectively.

3.2.3. Primary and Secondary Source Contributions

Chen *et al.* [31] found that fossil fuel combustion produces 80% ± 6% of EC in China. Therefore, EC is typically considered to be a tracer of primary organic carbon (POC). OC may originate directly from primary emissions generated by fuel combustion or indirectly from formation processes (e.g., secondary organic carbon, SOC) via gas-to-particle chemical reactions in the atmosphere [32]. However, distinguishing between primary organic carbon (POC) and SOC generated by chemical separation processes is difficult because there is no simple, direct analytical technique that can separate and quantify POC and SOC. In this study, the amount of SOC was estimated via the commonly used minimum OC/EC ratio method [17,25]. The OC/EC ratio is used in radiative transfer models to assess particle light scattering and absorption and to identify the presence of SOC. Ratios higher than 2.0 suggest the presence of SOC [33]. In our study, the OC levels were higher than the EC levels at all of the sampling sites; the OC/EC ratios ranged from 2.16 to 14.03. The relevant equations for this analysis are described as follows:

$$SOC = TOC - POC \tag{1}$$

$$POC = EC \times (OC/EC)_{min} \tag{2}$$

Where TOC is the total OC; $(OC/EC)_{min}$ is the minimum observed ratio.

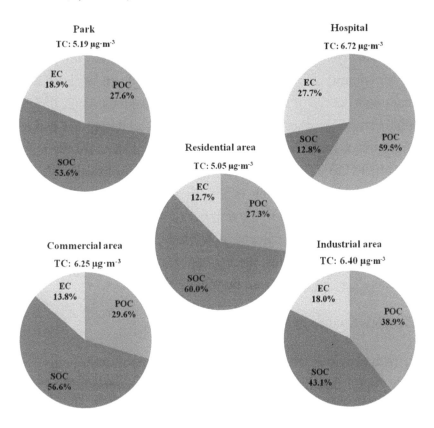

Figure 5. Relative contributions of primary and secondary sources to total carbon in the different functional areas in Xiamen.

Figure 5 shows the relative percentages of POC, SOC and EC in the observed total carbon (TC) within the different functional areas in Xiamen. SOC accounted for 53.6%, 56.6%, 43.1% and 60% of TC in parks, commercial areas, industrial areas and residential areas, respectively, and only 12.8% in hospitals. The results indicate that SOC formation is significant in these areas in Xiamen. The highest percentages of SOC in these areas are similar to the results of PM collected in southern cities of China during summer [25,26], which may be caused by the high potential photochemistry and low contributions from coal combustion. EC and POC accounted for 27.7% and 59.5% of TC at the hospitals. Sandrini *et al.* found that road-traffic influenced sites were characterized by the lowest levels of the OC/EC ratio, in the case with highest EC concentrations, denoting the prevalent primary origin of OC [28]. The highest percentage of POC and EC at the hospitals is primarily from motor vehicles. Hospitals in Xiamen are located in crowed areas and are common hubs for taxies, buses, and private cars. Exhaust emissions from the transport sector is a major source of EC. The statistical data show that vehicle usage has been rapidly increasing. Vehicle emissions have become a principal source of urban air pollution [34]. To reduce the impact of vehicle emissions on urban air quality, a series of vehicle emission controls and policies have been implemented, including the implementation of the European Union emission standards. However, legislation for vehicle emission control is focused on only a few compounds, such as carbon monoxide, total hydrocarbons, nitrogen oxides and particle numbers, with the exception of EC. Several studies have predicted that EC emissions from motor vehicles will be the largest and most important anthropogenic source in the coming decades [35]. The high values found at hospitals were primarily caused by vehicle emissions. For the convenience of patients, most of the hospitals are located along main roads, which typically have many vehicles. Thus the location is one of the reasons for the high EC values of observed at hospitals in Xiamen. Additionally, most hospitals do not have sufficient underground parking, and surface parking lots are in the vicinity of the sickrooms. Emissions from vehicles entering and exiting the parking lots were another major source of $PM_{2.5}$, OC, and EC at hospitals. Thus, limiting the numbers of vehicles entering the hospital campuses, constructing large-capacity underground parking structures, and locating the hospitals far from major roads in urban areas are essential steps for the long-term plans. The slightly higher contributions of EC to TC in the parks may also be attributed to the exhaust emissions from the transport sector. Many parks on Xiamen Island are not large and are commonly located near main roads. Furthermore, due to the increases in the number of vehicles in Xiamen, the limited number car parks cannot accommodate the increased motor vehicle usage; therefore, large empty areas of certain parks have been converted to parking lots. Thus, the EC concentrations in parks were slightly higher in Xiamen. Finally, the results may also be influenced by the limited dataset.

3.2.4. Comparison with Previous Studies in Other Areas

We compared our data with previous studies conducted in Xiamen, other areas in China, and other countries (see Table 7). We found that the $PM_{2.5}$, OC, and EC concentrations in Xiamen were much lower than those in Beijing [36] and Guangzhou [26], which provides clear evidence that Xiamen is one of the cleanest cities in China. Lower carbonaceous concentrations occurred in Xiamen, which is consistent with the low industrial coal consumption. However, the $PM_{2.5}$ concentrations in Xiamen exhibited a pronounced increased from 2003 to 2012, which may be a reason for the increasing frequency of atmospheric haze episodes in this region. Recent increase in PM levels may have been caused by increased motor vehicle usage, urban construction, and industrial combustion. The increase in traffic volume is particularly compelling. In Xiamen, the vehicle population increased from 275,000 in 2003 to 875,000 in 2012. However, due to the different analytical methods for OC and EC and different specific sampling sites, the OC and EC concentrations did not correspond with the $PM_{2.5}$. Previous studies have also observed that traffic-related emissions contribute to airborne OC and EC. Cyrys *et al.* [23] and Samara *et al.* [37] found that $PM_{2.5}$, OC and EC concentrations were obviously higher at the urban traffic sites than urban background areas. Sandrini *et al.* [28] found that OC concentrations increased by an order of magnitude from remote to traffic influenced sites and that EC concentrations increased by more than 50 times. The OC and EC concentrations in Xiamen were comparable to urban background sites in Europe, whereas the $PM_{2.5}$ concentrations were nearly three times higher than those in the European sites. Table 7 shows that the OC and EC concentrations in Xiamen were similar to those in Bari, which is a coastal site in Italy [28]. The similarity between these two sites may be attributed to their similarly coastal geographic positions.

Table 7. Comparison with data from previous studies.

Location	Site Type	Sampling Period	PM$_{2.5}$ (µg·m^{-3})	OC (µg·m^{-3})	EC (µg·m^{-3})	References
Xiamen, China	Urban	August–September 2012	43.70 ± 15.23	4.79 ± 1.19	1.09 ± 0.55	This study
Xiamen, China	Urban	June–July 2003	25.2 ± 15.8	4.8 ± 2.4	1.4 ± 1.3	Cao et al., 2007 [25]
Xiamen, China	Urban	April 2008 to February 2009	53.4 ± 25.0	7.6 ± 4.3	2.4 ± 0.8	Chen et al., 2012 [20]
	Rural	April 2008 to January 2011		5.7 ± 3.1	1.3 ± 0.7	
Xiamen, China	Urban	July 2009	34.26 ± 5.17	9.90 ± 0.67	2.34 ± 0.52	Zhang et al., 2011 [18]
Xiamen, China	Urban	August 2011	40.81 ± 27.89	6.87 ± 3.36	1.46 ± 0.80	Niu et al., 2013 [17]
Beijing, China	Urban	June–July 2009	73.8	12.5	4.9	Liu et al., 2014 [36]
Guangzhou, China	Urban	2006–2007 winter	-	8.53	4.81	Huang et al., 2012 [26]
		2006–2007 summer	-	5.97	3.46	
Germany	Urban background	March 1999 and July 2000	13.3	-	2.1	
	Urban traffic		14.3	-	3.1	
The Netherlands	Urban background	March 1999 and April 2000	17.8	-	2.1	Cyrys et al., 2003 [23]
	Urban traffic		19.9	-	3.9	
Sweden	Urban background	February 1999 and March 2000	10.2	-	1.4	
	Urban traffic		13.8	-	2.5	
Thessaloniki, Greece	Urban background	July–September 2011	23.5	5.72	0.69	Samara et al., 2014 [37]
	Urban traffic	February–April 2012	31.2	8.44	5.29	
San Pietro Capofiume, Italy	Rural	May–July 2007	-	3.1	0.78	
Bari Pane e Pomodoro, Italy	Urban background	March–July 2007	-	4.5	1.7	
Bari San Nicola, Italy	Urban background	March–July 2007	-	4.4	1.7	
Bari Casamassima, Italy	Urban background	March–July 2007	-	5.8	1.7	Sandrini et al., 2014 [28]
Milano Lodi, Italy	Urban background	February 2005 to June 2007		6.3	1.7	
Milano via Messina, Italy	Urban background	August 2002 to December 2003		9.3	1.4	
Milano Torre Sarca, Italy	Traffic	July 2005 to March 2008		9.4	4.5	

4. Conclusions

Intensive sampling of atmospheric PM (*i.e.*, TSP, PM_{10}, PM_5 and $PM_{2.5}$) was conducted from August to September in 2012 to provide information on fine-scale spatial variations in PM exposure in the urban area of Xiamen. The results showed that most TSP particles were inhalable particles of PM_{10} and that $PM_{2.5}$ accounted for more than 52.8% of the total PM_{10}. In terms of the average PM mass concentrations for all size categories, the order of the five functional area studies was as follows: hospital > park > commercial area > residential area > industrial area. OC, EC, and the major anions (*i.e.*, Cl^-, SO_4^{2-}, and NO_3^-) in $PM_{2.5}$ were also determined. OC contributed approximately 5%–23% of the total mass of $PM_{2.5}$, whereas EC only accounted for 0.8%–6.95%. High observed OC/EC ratios indicated the formation of SOC. Due to the high potential photochemistry in summer, SOC was the major carbonaceous particle in the parks, commercial areas, industrial areas, and residential areas, (*i.e.*, not at the hospitals). The highest percentages of POC and EC, observed at the hospitals, can be primarily attributed to motor vehicle sources. Although the results may be influenced by the limited dataset, this study has provided important insights into short-term PM exposures in the different functional areas in the urban area of Xiamen. Although the PM concentrations in Xiamen are much lower than in other cities in China, the concentrations appear to have increased significantly in recent years. Traffic vehicle emissions are a major source of the increased PM levels. Thus, it is essential to limit the numbers of vehicles in the urban area, especially near public areas, such as hospitals and parks. For the long-term future, properly locating hospitals, parks, and residential areas far from the main roads is critical.

Acknowledgements

This research was supported by the National Natural Science Foundation of China (41305133), Natural Science Foundation of Fujian Province (2013J05065), Special Fund for Marine Researches in the Public Interest (2004DIB5J178), and the open fund of Key Lab of Global Change and Marine-Atmospheric Chemistry of State Oceanic Administration (GCMAC1108). Finally, we would like to thank the reviewers and editors who have contributed valuable comments to improve the manuscript.

Author Contributions

The work presented here was carried out in collaboration with all authors. Liqi Chen and Ke Du supervised the research. Shuhui Zhao, Yanli Li, and Zhenyu Xing performed the field campaign. Shuhui Zhao involved the data analysis and wrote the manuscript. Liqi Chen and Ke Du contributed to the reviewing and revising of the manuscript.

Conflicts of Interest

The authors declare no conflict of interest.

References

1. Barima, Y.S.S.; Angaman, D.M.; N'Gouran, K.P.' Koffi, N.G.A.; Kardel, F.; De Cannière, C.; Samson, R. Assessing atmospheric particulate matter distribution based on Saturation Isothermal Remanent Magnetization of herbaceous and tree leaves in a tropical urban environment. *Sci. Total Environ.* **2014**, *470–471*, 975–982.

2. Watson, J.G. Visibility: Science and regulation. *J. Air Waste Manage. Assoc.* **2002**, *52*, 628–713.

3. Ramanathan, V.; Crutzen, P.J. New directions: Atmospheric brown clouds. *Atmos. Environ.* **2003**, *37*, 4033–4035.

4. Ulrich, P. Atmospheric aerosols: Composition, transformation, climate and health effects. *Angew. Chem. Int. Edit.* **2005**, *44*, 7520–7540.

5. Pope, C.A.; Dockery, D.W. Health effects of fine particulate air pollution: Lines that connect. *J. Air Waste Manage. Assoc.* **2006**, *56*, 709–742.

6. Katsouyanni, K.; Touloumi, G.; Samoli, E.; Gryparis, A.; Le Tertre, A.; Monopolis, Y.; Rossi, G.; Zmirou, D.; Ballester, F.; Boumghar, A.; *et al.* Confounding and effect modification in the short-term effects of ambient particles on total mortality: Results from 29 European cities within the APHEA2 project. *Epidemiology* **2001**, *12*, 521–531.

7. Samet, J.M.; Zeger, S.L.; Dominici, F.; Curriero, F.; Coursac, I.; Dockery, D.W.; Schwartz, J.; Zanobetti, A. The National morbidity, mortality, and air pollution study. Part II: Morbidity and mortality from air pollution in the United States. *Res. Rep. Health. Eff. Inst.* **2000**, *94*, 5–70.

8. World Health Organization (WHO). *Air Quality Guidelines for Particulate Matter, Ozone, Nitrogen Dioxide and Sulfur Dioxide,Global Update 2005, Summary of Risk Assessment*; WHO: Geneva, Switzerland, 2006.

9. Raaschou-Nielsen, O.; Andersen, Z.J.; Beelen, R.; Samoli, E.; Stafoggia, M.; Weinmayr, G. Air pollution and lung cancer incidence in 17 European cohorts: Prospective analyses from the European Study of Cohorts for Air Pollution Effects (ESCAPE). *Lancet. Oncol.* **2013**, *14*, 813–822.

10. Che, H.; Zhang, X.; Li, Y.; Zhou, Z.; Qu, J.J.; Hao, X. Haze trends over the capital cities of 31 provinces in China, 1981–2005. *Theor. Appl. Climatol.* **2009**, *97*, 235–242.

11. Zhao, P.; Zhang, X.; Xu, X.; Zhao, X. Longterm visibility trends and characteristics in the region of Beijing, Tianjin, and Hebei, China. *Atmos. Res.* **2011**, *101*, 711–718.

12. Monn, C.; Carabias, V.; Junker, M.; Waeber, R.; Karrer, M.; Wanner, H.U. Small-scale spatial variability of particulate matter <10 mu m (PM$_{10}$) and nitrogen dioxide. *Atmos. Environ.* **1997**, *31*, 2243–2247.

13. Lung, S.C.C.; Mao, I.F.; Liu, L.J.S. Residents' particle exposures in six different communities in Taiwan. *Sci. Total Environ.* **2007**, *377*, 81–92.

14. Lung, S.-C.C.; Hsiao, P.-K.; Wen, T.-Y.; Liu, C.-H.; Fu, C.B.; Cheng, Y.-T. Variability of intra-urban exposure to particulate matter and CO from Asian-type community pollution sources. *Atmos. Environ.* **2014**, *83*, 6–13.

15. Fan, X.; Sun, Z. Analysis on features of haze weather in Xiamen City during 1953–2008. *Trans. Atmos. Sci.* **2009**, *32*, 604–609. (In Chinese)

16. Niu, Z.; Zhang, F.; Kong, X.; Chen, J.; Yin, L.; Xu, L. One-year measurement of organic and elemental carbon in size-segregated atmospheric aerosol at a coastal and suburban site in Southeast China. *J. Environ. Monit.* **2012**, *14*, 2961–2967.

17. Niu, Z.; Wang, S.; Chen, J.; Zhang, F.; Chen, X.; He, C.; Lin, L.; Yin, L.; Xu, L. Source contributions to carbonaceous species in $PM_{2.5}$ and their uncertainty analysis at typical urban, peri-urban and background sites in southeast China. *Environ. Pollut.* **2013**, *181*, 107–114.

18. Zhang, F.; Zhao, J.; Chen, J.; Xu, Y.; Xu, L. Pollution characteristics of organic and elemental carbon in PM2.5 in Xiamen, China. *J. Environ. Sci.* **2011**, *23*, 1342–1349.

19. Zhang, F.; Xu, L.; Chen, J.; Yu, Y.; Niu, Z.; Yin, L. Chemical compositions and extinction coefficients of $PM_{2.5}$ in peri-urban of Xiamen, China, during June 2009–May 2010. *Atmos. Res.* **2012**, *106*, 150–158.

20. Chen, B.; Du, K.; Wang, Y.; Chen, J.; Zhao, J.; Wang, K.; Zhang, F.; Xu, L. Emission and transport of carbonaceous aerosols in urbanized coastal areas in China. *Aerosol Air Qual. Res.* **2012**, *12*, 371–378.

21. Weather History & Data Archive from Weather Underground, the Weather Channel, Inc. Available online: http://www.wunderground.com (accessed on 27 January 2015).

22. Birch, M.E.; Cary, R.A. Elemental carbon-based method for monitoring occupational exposures to particulate diesel exhaust: Methodology and exposure issues. *Aerosol Sci. Tech.* **1996**, *121*, 1183–1190.

23. Cyrys, J.; Heinrich, J.; Hoek, G.; Meliefste, K.; Lewne, M.; Gehring, U.; Bellander, T.; Fischer, P., Van Vliet, P.; Brauer, M.; *et al.* Comparison between different traffic-related particle indicators: Elemental. carbon (EC), $PM_{2.5}$ mass, and absorbance. *J. Expo. Anal. Environ. Epidemiol.* **2003**, *13*, 134–143.

24. Holgate, S.T.; Samet, J.M.; Koren, H.S.; Maynard, R.L. *Air Pollution and Health*; Academic Press: London, UK, 1999.

25. Cao, J.J.; Lee, S.C.; Chow, J.C.; Watson, J.G.; Ho, K.F.; Zhang, R.J.; Jin, Z.D.; Shen, Z.X., Chen, G.C.; Kang, Y.M.; *et al.* Spatial and seasonal distributions of carbonaceous aerosols over China. *J. Geophys. Res.* **2007**, doi:10.1029/2006JD008205.

26. Huang, H.; Ho, K.F.; Lee, S.C.; Tsang, P.K.; Ho, S.S.H.; Zou, C.W.; Zou, S.C.; Cao, J.J.; Xu, H.M. Characteristics of carbonaceous aerosol in $PM_{2.5}$: Pearl Delta River Region, China. *Atmos. Res.* **2012**, *104–105*, 227–236.

27. Turpin, B.J.; Lim, H.J. Species contributions to $PM_{2.5}$ mass concentrations: Revisting common assumptions for estimating organic mass. *Aerosol Sci. Tech.* **2001**, *35*, 602–610.

28. Sandrini, S.; Fuzzi, S.; Piazzalunga, A.; Prati, P.; Bonasoni, P.; Cavalli, F.; Bove, M.C.; Calvello, M.; Cappelletti, D.; Colombi, C.; *et al.* Spatial and seasonal variability of carbonaceous aerosol across Italy. *Atmos. Environ.* **2014**, *99*, 587–598.

29. Contini, D.; Cesari, D.; Donateo, A.; Chirizzi, D.; Belosi, F. Characterization of PM_{10} and $PM_{2.5}$ and their metals content in different typologies of sites in South-Eastern Italy. *Atmosphere* **2014**, *5*, 435–453.

30. Putaud, J.P.; van Dingenen, R.; Alastuey, A.; Bauer, H.; Birmili, W.; Cyrys, J.; Flentje, H.; Fuzzi, S.; Gehrig, R.; Hansson, H.C.; *et al.* A European aerosol phenomenology-3: Physical and chemical characteristics of particulate matter from 60 rural, urban, and kerbside sites across Europe. *Atmos. Environ.* **2010**, *44*, 1308–1320.

31. Chen, B.; Andersson, A.; Lee, M.; Kirillova, E.N.; Xiao, Q.F.; Krusa, M.; Shi, M.N.; Hu, K.; Lu, Z.F.; Streets, D.G.; *et al.* Source forensics of black carbon aerosols from China. *Environ. Sci. Technol.* **2013**, *47*, 9102–9108.

32. Seinfeld, J.H.; Pandis, S.N. *Atmospheric Chemistry and Physics: From Air Pollution to Climate Change*; John Wiley: New York, NY, USA, 1998.

33. Chow, J.C.; Watson, J.G.; Lu, Z.; Lowenthal, D.H.; Frazier, C.A.; Solomon, P.A.; Thuillier, R.H.; Magliano, K. Descriptive analysis of $PM_{2.5}$ and PM_{10} at regionally representative locations during SJVAQS/AUSPEX. *Atmos. Environ.* **1996**, *30*, 2079–2112.

34. Bond, T. Can warming particles enter global climate discussions? *Environ. Res. Lett.* **2007**, doi:10.1088/1748-9326/2/4/045030.

35. Streets, D.G.; Bond, T.C.; Lee, T.; Lang, C. On the future of carbonaceous aerosol emissions. *J. Geophys. Res.* **2004**, doi:10.1029/2004JD004902.

36. Liu, Y.J.; Zhang, T.T.; Liu, Q.Y.; Zhang, R.J.; Sun, Z.Q.; Zhang, M.G. Seasonal variation of physical and chemical properties in TSP, PM_{10} and $PM_{2.5}$ at a roadside site in Beijing and their influence on atmospheric visibility. *Aerosol Air Qual. Res.* **2014**, *14*, 954–969.

37. Samara, C.; Voutsa, D.; Kouras, A.; Eleftheriadis, K.; Maggos, T.; Saraga, D.; Petrakakis, M. Organic and elemental carbon associated to PM_{10} and $PM_{2.5}$ at urban sites of northern Greece. *Environ. Sci. Pollut. Res.* **2014**, *21*, 1769–1785.

9

Effects of Aerosol on Cloud Liquid Water Path: Statistical Method a Potential Source for Divergence in Past Observation Based Correlative Studies

Ousmane Sy Savane [1,*], **Brian Vant-Hull** [2], **Shayesteh Mahani** [2] and **Reza Khanbilvardi** [2]

[1] Earth and Environmental Sciences Department, Graduate Center, The City University of New York, New York, NY 10031, USA

[2] Cooperative Remote Sensing Science and Technology (CREST) Institute, City University of New York, New York, NY 10031, USA; E-Mails: brianvh@ce.ccny.cuny.edu (B.V-H.); mahani@ce.ccny.cuny.edu (S.M.); rk@ce.ccny.cuny.edu (R.K.)

* Author to whom correspondence should be addressed; E-Mail: Osysavane8@gmail.com

Academic Editor: Toshihiko Takemura

Abstract: Studies show a divergence in correlation between aerosol and cloud proxies, which has been thought of in the past as the results of varying physical mechanisms. Though modeling studies have supported this idea, from an observational standpoint it is difficult to attribute with confidence the correlations to specific physical mechanisms. We explore a methodology to assess the correlation between cloud water path and aerosol optical depth using Moderate-resolution Imaging Spectroradiometer (MODIS) Aqua retrieved aerosol and cloud properties for absorbing and non-absorbing aerosol types over land and over the Atlantic Ocean for various meteorological conditions. The data covers a three-month period, June through August, during which different aerosol types are predominant in specific regions. Our approach eliminates outliers; sorts the data into aerosol bins; and the mean Aerosol Optical Depth (AOD) value for each bin and the corresponding mean Cloud Water Path (CWP) value are determined. The mean CWP is plotted against the mean AOD. The response curve for all aerosol types shows a peak CWP value corresponding to an aerosol loading value AOD_{peak}. The peak is used to divide the total range of aerosol loading into two sub ranges. For AOD value below AOD_{peak}, mean CWP and mean AOD are positively correlated. The correlation between mean CWP and mean AOD is negative for aerosol loading above AOD_{peak}. Irrespective of aerosol type, atmospheric water vapor content and lower tropospheric static stability, the peak observed for each aerosol type seems to describe a universal feature that calls for further investigation. It has been observed for a variety of geographical locations and different seasons.

Keywords: cloud water path (CWP); aerosol optical depth (AOD); correlation

---

List of Symbols

AOD	Aerosol Optical Depth
LTSS	Lower Tropospheric Static Stability
WV	Water Vapor
CWP	Cloud Water Path
AAI	Absorbing Aerosol Index
NAADR_Marine	Non Absorbing Aerosol Dominated Region in the marine environment. This aerosol type has a strong sea salt component.
NAADR_Land	Non Absorbing Aerosol Dominated Region over land. This aerosol type has a strong sulfates component.
AADR_Sahara	Absorbing Aerosol Dominated Region in the Sahara environment. This aerosol type has a strong dust component.
AADR_Urban	Absorbing Aerosol Dominated Region in urban environment. This aerosol type has a strong soot component.
AADR_SubTrop	Absorbing Aerosol Dominated Region in Subtropical African biomass burning region. This aerosol type has a strong smoke component

1. Introduction

In the context of global climate change, aerosol and cloud interactions remain the most uncertain factor because the mechanisms that link aerosol to cloud properties change are carried out through processes that are not well understood [1]. Investigations involving cloud and aerosol showed positive, negative or no correlation between aerosol and cloud properties in general and cloud water path particularly. The Advanced Very High Resolution Spectro-Radiometer (AVHRR) retrieved cloud and aerosol properties showed a negative correlation between the column cloud condensation nuclei (CCN) concentration and the effective radius and a positive correlation with cloud liquid water path over the ocean [2]. The POLarization and Directionality of the Earth's Reflectance (POLDER) instrument retrieved cloud and aerosol properties that showed a negative correlation exists over both land and ocean [3]. The AVHRR water path and the column CCN were found not only positively correlated [4] but also negatively correlated on the global scale [5]. It has been suggested that the correlation between these parameters may be positive or negative and vary regionally [6], while the results of a study conducted by Kaufman in 2005 suggested that the sign of the correlation may depend on aerosol type [7]. One of the most significant consequences of the observed divergences in correlative studies was the intense search for possible physical mechanisms or factors that could explain the disparities in the results between scientists.

1.1. Suggested Mechanisms Leading a Positive Correlation

The mechanisms likely to result in increased in cloud liquid water as aerosol loading increases are associated with precipitation.

a. Aerosol affects clouds: The addition of aerosol causes a decrease in drop size, precipitation is suppressed, and clouds develop further before raining out (if they ever do) and last longer in the more developed stage, thus increasing average Cloud Water Path (CWP) [8,9].

b. Clouds affect aerosol: Following a precipitation event the aerosol loading is dramatically reduced, as is the cloud development (the clouds "rain out"). Low aerosol is thus associated with low CWP [10].

1.2. Suggested Mechanisms Leading a Negative Correlation

Several mechanisms would cause cloud development to decrease as aerosol loading increases.

a. Aerosol affects clouds: (Surface shading) Aerosols shade the surface, reducing surface heating and evapotranspiration so that cloud liquid water is reduced [11].

b. Aerosol affects clouds: (Atmospheric heating) Absorbing aerosols (such as smoke or dust) can heat the upper levels of the troposphere, which in combination with surface shading stabilizes the atmospheric column and reduces cloud development [11–14].

c. Aerosol affects clouds: (Clouds drop size) As an increase in CCN leads to smaller droplets, evaporation around the sides and top of clouds due to mixing will become more effective at reducing the cloud liquid water [11,15].

d. Aerosol and meteorology affect clouds: High-pressure systems inhibit convective activity, simultaneously reducing cloudiness while not allowing smoke (other aerosols with sources in the region) to "vent" away from the source region [16].

e. Clouds and meteorology affect aerosol: In a humid environment and in areas surrounding clouds, significant aerosol hygroscopic growth has been observed [17].

1.3. Other Factors

Many studies pointed out the methodological differences in assessing the effects of aerosol on CWP as possible source for divergence.

a. Bulk scheme simulation *vs.* Bins microphysics scheme simulation [18,19].

b. CWP averaging method over the domain: full grid points averaging *vs.* conditional (cloud only) grid points averaging [20].

c. Cloud regimes: Stratocumulus *vs.* Trade wind cumulus [21].

d. Analysis scales *vs.* Process scales [22].

Exploring alternative explanations to understand the disparity between studies is the motivation behind this study. Here, we focus on the statistical method used to trace the relationship between observational aerosol and cloud proxies, which, to our knowledge, has not been questioned in these past studies. Correlations in many studies have been established by linearly regressing aerosol proxies against cloud properties (as indicated in Figure 1). An example of this approach could be found in [23].

However, the results of statistics applied to a dataset and the inferred interpretations could vary greatly from one range of the dataset to another. In the context of cloud and aerosol interaction, modeling studies in a controlled environment can discretely increase aerosol loading and for each aerosol loading determine precisely the modeled the response of cloud proxies. The results of these modeling studies have shown a strong correlation between aerosol loading range and the response of cloud water path [24,25]. Unlike modeling studies, observational studies rely on snapshots of the atmospheric scenes produced by satellites to infer correlations between aerosols optical depth and cloud water path, requiring assumptions about the cloud lifecycle to relate correlations to physical mechanisms [6,10,26]. In addition, many observational studies in the past did not consider the effects of varying Aerosol Optical Depth (AOD) ranges in their statistical analysis in order to determine the correlation between CWP and AOD. In Figure 2 for example, the mean CWP response to the mean AOD varies in terms of mean AOD ranges. Which correlation describes best the relationship between aerosol and cloud proxies in the dataset?

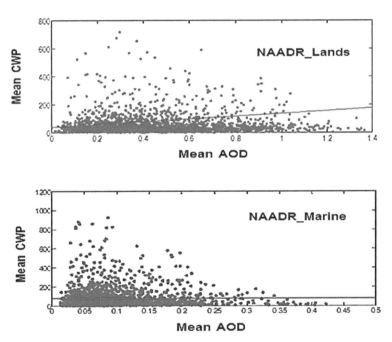

Figure 1. Shows for both NAADR_Land (sulfate) and NAADR_Marine (sea salt), respectively, a positive and a negative correlations by single line linear regression method between mean cloud water path (CWP) and mean aerosol optical depth (AOD) over the total AOD range, for non-binned aerosol data and outliers not removed. In contrast to techniques presented later, there is no averaging or removal of outliers.

The methodology we propose characterizes the response of cloud water path in terms of aerosol loading ranges for satellite data. Our results could shed some light on a possible source of the disparities in observation based past correlative studies.

This paper is organized around six major Sections. Section 1 presents the context of this work. Section 2 describes both the study areas where investigations were conducted as well as the data used. The steps of the analytical method and the partial results when the method is applied to NAADR_Marine and NAADR_Land are described in Section 3. Section 4 evaluates the effects of meteorology on CWP response to increasing AOD by both statistic compositing and multilinear regression analysis. Section 5 contains the discussion part of the paper. In this Section, a parallel is established between the single line linear regression method (Figure 1) and our approach in order to reveal a potential source of divergence in past correlative studies. Finally Section 6 concludes the paper by summarizing the main points we attempted to make.

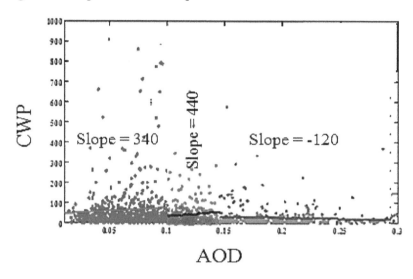

Figure 2. Correlations between aerosol optical depth (AOD) and cloud water path (CWP) in southern region of Africa for three different ranges of AOD from the same dataset. The low ranges of AOD (0; 0.1) (blue data points) and (0.1; 0.15) (red data points) show a positive correlation between AOD and CWP. The highest range of AOD (0.15; 0.3) (green data points) shows a negative correlation between AOD and CWP for (0.15; 0.3). The ranges were obtained by gradually increasing by 0.05 (arbitrary constant) AOD loading in the scatter plot CWP *vs.* AOD until a significant change (magnitude or sign) in the slope of the regression line occurs.

2. Study Area and Satellite Data

2.1. Study Area

During June through August, the Atlantic Ocean is covered by varying concentrations of several aerosol types, with each type dominant in a distinct latitude belt [7] as indicated in Figure 3. The regions of interest are identified by either absorbing or non-absorbing aerosol type. Sampled in Northern Atlantic (30°N–60°N), study region (**B**) is impacted by anthropogenic pollution aerosol (Sulfate) from North America and Europe. Study Area (**A**) is sampled in the Southern Tropical Atlantic (30°S–20°S) which is under strong influence of clean maritime air (Sea salt) [7]. The areas (**C**) and (**D**) are dominated by absorbing aerosol types as they are, respectively, sampled from Sahara dust and Sub-Tropical biomass burning African regions. Study region (**E**) is sampled in an area of high urban impact and is absorbing aerosol type dominated (Soot). A study area located over ocean and dominated by absorbing aerosol will be marked as AADR_Marine ("Absorbing Aerosol Dominated Region") and non-absorbing aerosol, will be marked as NAADR_Marine ("Non Absorbing Aerosol Dominated Region"). For study areas, respectively, sampled in the Sahara dust and Sub-Tropical smoky background environments, the regions are marked as AADR_Sahara (**C**) and AADR_SubTrop (**D**). In addition, summertime in the Northern Hemisphere corresponds to wintertime in Southern Hemisphere. Study periods for E, B and C in the Northern Hemisphere covered June through August (summer) for A and D this will correspond to wintertime in the Southern Hemisphere.

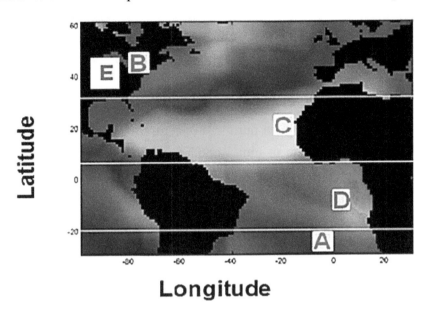

Figure 3. Shows the latitudinal distribution of study areas. Regions (**A**)–(**E**) are, respectively, characterized by as NAADR_Marine, NAADR_Land, AADR_Sahara, AADR_SubTrop, and AADR_Urban.

2.2. Satellite Data

The data used is MODIS joint atmospheric product level 2 retrieved from June through August 2005 between 14:00 and 22:00 GMT on board of Aqua platform. The day time Aqua overpass (13:30) coverage is chosen because clouds are more likely to develop in the afternoon than in the morning [26]. Once retrieved, the data was processed to meet the conditions of this study; we want to study aerosol and cloud interaction from a liquid cumulus clouds perspective because retrieval errors are smaller in the pure liquid phase, and spaces between cumulus clouds allow retrieval of aerosol nearby. The cloud top temperature (CTT) and cloud fraction (CF) were, respectively, maintained above 265 °K and less than 0.6 to capture pixels of warm, liquid cumulus clouds. The highest data density was produced by solar zenith angles between 30 and 70 degrees and view angles less than 60 degrees. The data was filtered according to these restricted angular ranges in order to minimize the influence of biases due to three-dimensional effects [10].

3. Method

Examples of the analytical method are presented using two study areas NAADR_Land and NAADR_Marine. Responses of CWP to aerosol induced perturbation for the remaining study areas are displayed with the two initial study areas in Section 3 to exhibit the general nature of the results.

3.1. Mathematical Relationships and Definition

(a) Liquid Water Sensitivity (δ)

The concepts of liquid water sensitivity and relative liquid water sensitivity were used by Han *et al.* [6] and we use the same definition with a single change in variable to assess the response of cloud water path to aerosol induced perturbation. Liquid water sensitivity as defined in Equation (3) represents the change of liquid water path *vs.* column droplet number concentration (a proxy for aerosol particle count), which is affected by the total water availability [6]. The sensitivity is denoted by δ, and is derived using the least-squares linear regression to determine the slope of ΔLWP (change in cloud liquid water path) and ΔN_c (change in droplets number concentration). This formulation of δ assesses the response of cloud water path response to cloud condensation nuclei independently of the actual aerosol type retrieved. This was intentional to eliminate dependence on aerosol type.

The two variables used to compute δ, are derived from both cloud drop effective radius (r_e) and cloud optical depth(τ). The droplet number of concentration N_c is defined as function of CWP [27] and represented by Equation (1).

$$N_c = \frac{3}{4\pi\rho_w} \cdot \frac{\text{CWP}}{r_e^3(1-b)(1-2b)} \tag{1}$$

where r_e is the cloud drop effective radius and b is effective variance of cloud droplet size distribution. CWP is calculated using both τ and r_e [28] as indicated in Equation (2).

$$\text{CWP} = \frac{2}{3}r_e\tau\rho_w \tag{2}$$

where τ is the cloud optical depth.
From these two definitions, δ is derived [6] as indicated in Equation (3).

$$\delta = \frac{\Delta\text{CWP}}{\Delta N_c} \tag{3}$$

(b) Liquid Water Relative Sensitivity β

When the liquid water sensitivity is normalized for different environments, it isolates better the effect of aerosol-cloud interaction [6]. Cloud water path relative sensitivity is defined as in Equation (4).

$$\beta = \frac{\Delta\text{CWP}/\text{CWP}}{\Delta N_c / N_c} \approx \frac{\Delta\ln(\text{CWP})}{\Delta\ln(N_c)} \tag{4}$$

Approximating the droplet number of concentration (sometimes referred to as aerosol loading in this study) to aerosol optical depth [29], we define the relative sensitivity β in terms of CWP and AOD in our study. β is then used to statistically trace aerosol-cloud relationship. This approximation is based on the assumption that moisture is uniformly distributed in the cloudy atmosphere. In that instance, aerosol swelling in humid environment could be interpreted as an increase in aerosol loading [17]. Increase in aerosol loading produces numerous but smaller cloud droplets [30,31]. The presence of ratio in the formula of the relative sensitivity compensates for possible variations in the proportionality factor between N_c and AOD, and similar variations in retrievals of CWP. This makes β a suitable parameter to assess the correlation between CWP and AOD. We assumed $N_c = k \times$ AOD where k is a constant, then k will cancel out in Equation (4), so that our form of beta expressed in terms of CWP and AOD and represented by Equation (5) is the same as in Equation (4).

$$\delta = \Delta CWP \Big/ \Delta AOD$$

$$\beta = \Delta Log[CWP] \Big/ \Delta Log[AOD]$$

(5)

β sometime has been known in literature as λ and is believed to be difficult to measure because of rapid adjustment of CWP to aerosol-induced perturbation [20].

Much of the effect on CWP is due to rainfall, as seen by both modeling [32,33] and observations [34] studies. The picture remains unclear with variability of the results in the literature attributed to averaging scale [35], phases of cloud life cycle [35,36] and aerosol loading [36]. Given these factors plus the relative sparseness of retrieved rainfall over the ocean sites in this study, we chose to leave rainfall as a hidden variable in this study.

The goal here is not so much to accurately quantify β, but to assess its variability. We will demonstrate the non-monotonic behavior of the CWP response to increasing AOD by computing and comparing the slopes of CWP *vs.* AOD, log(CWP) *vs.* log(AOD) and the corresponding correlation coefficients during "moistening" and "drying" (these terms are clearly defined in Section 3 for different types of aerosol when: (1) the data is not binned and outliers are not removed; and (2) the data is averaged by bins and outliers are removed).

In addition to aerosol, cloud formation involves both the dynamics of convection and moisture availability. We anticipate the sensitivity of CWP response to atmospheric static stability as well the atmospheric content of precipitable water vapor as meteorological parameters. CWP response for each type of aerosol is determined for low and high Water Vapor (WV) environment as well as low and high lower tropospheric static stability (LTSS) environment in order to assess the influence of meteorology. Moreover, to determine which variables (aerosol or meteorology) is the governing variable of CWP response (increase/decrease during Moistening/Drying), CWP and all the meteorological parameters are binned, respectively, in AOD moistening and drying ranges and coupled with multilinear regression analysis in each case.

3.2. Data Preparation (Elimination of Outliers/Bins Averaging)

Outliers may occur in cloud and aerosol data for many reasons. Cloud contamination may affect aerosol data [7]. Sources of cloud contamination include:

1. High concentration of broken clouds that may induce illumination of aerosol field beyond 500-m distance from the clouds [37].

2. Relative humidity can affect both aerosol and cloud amount and may result in a change of the ratio of aerosol optical depth to cloud condensation nuclei through aerosol swelling [38].

This Section will be illustrated using two study areas. NAADR_Land located in Northern hemisphere and NAADR_Marine located in Southern Hemisphere. We will only show the results for the remaining study areas. Least squares regressions are strongly influenced by outliers. In order to avoid skewed correlations between CWP and AOD, we used the data smoothing technique according to Eilers *et al.*, [39] where outliers are defined as points regions where data density is less than a fraction of the maximum density. The minimum threshold density to deciding which data points to be considered for this study was set at 50, approximately 20% of the maximum density. We carefully select a box avoiding outliers (white dots) containing data points density layers ranging from 50 to 250 to define our working data ranges; the corresponding data ranges were: AOD (0.05–0.60), (0.01–0.20), respectively, for region **B** and for region **A**. CWP (0–200 g/m²) for both regions (see Figure 4). Many other boxes with data point's density ranges and defining various AOD and CWP ranges can be drawn; our choice maximized the dataset size by eye.

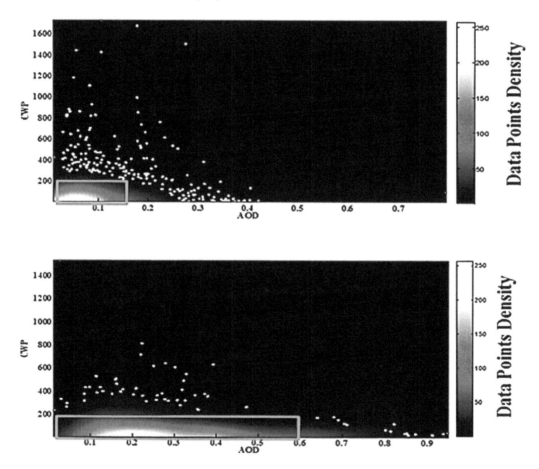

Figure 4. Data points' density distribution in a cloud water path (CWP) *vs.* aerosol optical depth (AOD) plot for NAADR_Marine (**Top**) and NAADR_Land (**Bottom**). The green rectangle indicates the best ranges of AOD and CWP to obtain a minimum of 50 data points density per AOD bin (0.01) and (0.05), respectively, for NAADR_Marine and NAADR_Land. The range for AOD are [0, 0.15] & [0, 0.6] and CWP [0; 200].

To demonstrate the non-monotonic behavior of CWP response, we divide the full range of AOD into bin sizes of 0.05 (NAADR_Land) and 0.01 (NAADR_Marine). The characteristics mean AOD and mean CWP per data points in each bin were calculated and Mean AOD is plotted against mean CWP (Figure 5).

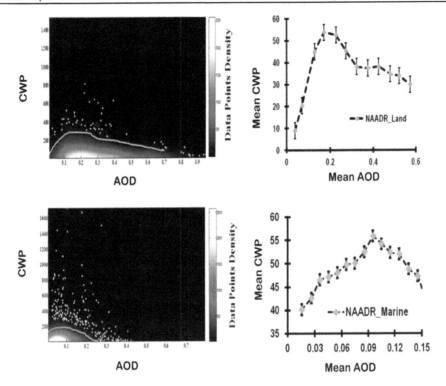

Figure 5. (Top) Shows for NAADR_Land the average behavior of cloud water path (CWP) to increasing aerosol optical depth (AOD) for all data density layers excluding outliers (White dots) (**Left**) and the plot of CWP *vs.* AOD after binning (**Right**). Each bin will have both an average and a standard deviation (**Bottom**) Shows for NAADR_Marine the average behavior of CWP to increasing AOD for all data density layers excluding outliers (White dots) (left) and the plot of CWP *vs.* AOD after binning (right).

3.3. Description of Cloud Water Path (CWP) Response to Increasing Aerosol Optical Depth (AOD)

When AOD mean is plotted against mean CWP, a peak is observed in both graphs. This peak is used to divide the full data range into two sub ranges. As aerosol loading increases, the graphs show an initial increase in CWP to the peak followed by a decrease for all five regions (see Figure 6). For simplification purposes and to indicate the occurrence of two distinct and opposite behaviors, the initial increase in CWP to maximum (CWP_{peak}) that occurs as aerosol loading increases is referred to as "moistening" and the decrease in CWP from CWP_{peak} that occurs as aerosol loading increases from AOD_{peak} to higher aerosol loading is referred to as "drying". "Moistening" and "drying" are descriptive of the shape of the response curves only, and are not meant to represent any physical processes that are actually occurring in these ranges (see Figure 7). The relative sensitivity of CWP to AOD β is calculated as the slope of the plot log (mean CWP) *vs.* log (mean AOD) for each sub-range. The magnitudes and the signs of β (see Table 1) show a non-monotonic response of cloud water path to increasing aerosol loading contrary to a monotonic response suggested when Single Line Linear Regression method was applied to the same datasets. The correlation coefficients for both binned non-logarithmic and binned logarithmic data in the two regions are significantly high compared to the correlation coefficients determined before the data is binned and the outliers are removed. A large part of the increase in correlation coefficients is due to averaging, but also to the excluded outliers' contributions.

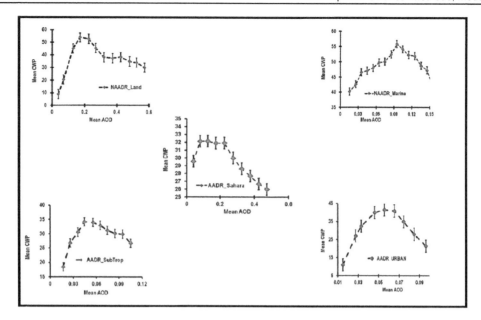

Figure 6. Responses of cloud water path (CWP) to increasing aerosol optical depth (AOD) for all five study regions. CWP initially increases to a maximum followed by a decrease in each case. CWP sensitivities during moistening and drying are calculated from the response curves as the slopes of mean CWP *vs.* mean AOD during both moistening and drying. For each average point shown above, the standard deviation is shown as error bars.

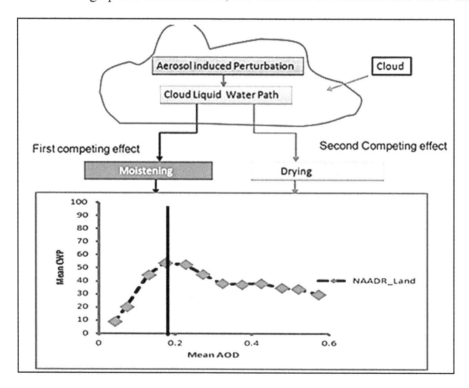

Figure 7. Illustration of cloud and aerosol interaction that resulted in an initial increase (moistening) of cloud water path (CWP) to a maximum followed by a decrease (drying) as aerosol optical depth (AOD) increases for NAADR_Land.

Table 1. Correlation coefficients (R^2), sensitivity (δ), relative sensitivity (β) and slope (a^*) of the regression line (before binning) of cloud water path (CWP) to increasing aerosol optical depth (AOD) as function of AOD ranges and aerosol type.

Parameters	R^2/β	R^2/δ	R^2/a^*
NAADR_Land	--	--	--
AOD Ranges	--	--	--
[0.02, 0.172]	$0.991/1.22 \pm 1.69$	$0.984/348.3 \pm 1.74$	$0.183/180 \pm 0.226$
[0.172, 0.6]	$0.916/-0.514 \pm 1.69$	$0.844/-53.54 \pm 1.74$	$0.05/-160 \pm 0.226$
NAADR_Marine	--	--	--
AOD Ranges	--	--	--
[0, 0.095]	$0.950/0.16 \pm 0.27$	$0.950/171.7 \pm 1.88$	$0.003/130 \pm 0.271$
[0.095, 0.15]	$0.860/-0.58 \pm 0.27$	$0.911/-221 \pm 1.88$	$0.0004/-270 \pm 0.271$

* Slope is computed before outliers are removed and the data is sorted into aerosol optical depth bins.

4. Effects of Varying Meteorological Parameters on Cloud Water Path (CWP) Response to Increasing Aerosol Optical Depth (AOD)

Although it is tempting to speculate on the physical mechanisms behind the observations produced by these two methods, the context in which the interactions occur between variables is too complex to be completely explored by the current dataset, so we prefer to leave such discussion to future investigations. The reader is referred to the many possible mechanisms described in the introduction.

4.1. Evaluation by Atmospheric Water Vapor (WV) Statistical Compositing

The responses of CWP to increasing AOD are evaluated in high and low atmospheric water vapor content environment for the five study areas. WV path was used as a direct satellite measurement. The total WV range for each study area is divided into two subranges of nearly equal counts based on WV histogram data distribution. Low WV range is low WV relative to high WV (High WV range).

The characteristics parameters of CWP response curves are determined in each case and tabulated for comparison purpose.

4.1.1. NAADR_Marine

In Figure 8, the threshold aerosol loading necessary to shift from cloud moistening process to drying process is larger in high water vapor environment than in low water vapor environment as well as the corresponding cloud water paths.

In Table 2, the transition from low to high water vapor regime resulted in a significant increase ($\uparrow$) in keys parameters (AOD_{peak}, CWP_{peak}, $\beta_{moistening}$, β_{drying}) of cloud water path response curve to aerosol induced perturbation. β_{drying} increases nearly 10 times from a low to high water vapor regime and this is also the most significant magnitude change. The lowest magnitude change occurred in $\beta_{moistening}$ (29%).

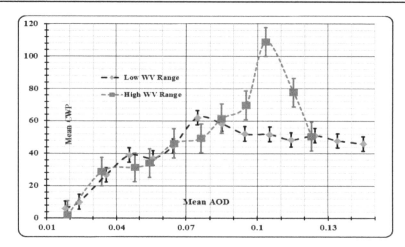

Figure 8. Response of cloud water path (CWP) to increasing aerosol optical depth (AOD) in low and high atmospheric WV environments for NAADR_Marine.

Table 2. Comparative table of the percent increase (↑) in the characteristics parameters of cloud water path response curve to aerosol induced perturbation from low to high water vapor environment for NAADR_Marine.

Moisture	AOD_{peak}	CWP_{peak}	$\beta_{moistening}$	β_{drying}
Low WV	0.074 ± 0.011	62.13 ± 1.74	1.59 ± 0.17	-0.417 ± 0.17
High WV	0.104 ± 0.011	109 ± 1.74	2.050 ± 0.20	-4.319 ± 0.20
($\Delta X/X$ Low WV) %	40.5% ↑	75.4% ↑	29% ↑	936% ↑

4.1.2. NAADR_Lands

The threshold aerosol loading necessary to shift from cloud moistening process to drying process is much larger in low water vapor environments than in high water vapor environments as well as the corresponding cloud water paths (Figure 9). The availability of moisture in high WV environments could offset the drying effect induced by increasing aerosol loading compared to CWP response in low water vapor environments. In Table 3, except for $\beta_{moistening}$, all characteristic parameters AOD_{peak}, CWP_{peak}, β_{drying} for NAADR_ Land decreased significantly from low to high water vapor environments In high WV vapor environments, $\beta_{moistening}$ exhibits the highest magnitude compared to β_{drying} (moistening is the highest sensitivity mechanism) while drying the is the highest sensitivity mechanism in low WV conditions. During the transition from low to high atmospheric water vapor content, the most significant magnitude change occurred in $\beta_{moistening}$ (143% increase) while the lowest magnitude change occurred in CWP_{peak} (11% decrease).

Table 3. Comparison of percent increase (↑) or decrease (↓) in the characteristics parameters of cloud water path response curve aerosol induced perturbation from low to high water vapor environment NAADR_Land.

Moisture	AOD_{peak}	CWP_{peak}	$\beta_{moistening}$	β_{drying}
Low WV	0.419 ± 0.036	77.40 ± 2.84	0.212 ± 0.06	-0.492 ± 0.180
High WV	0.275 ± 0.036	69.11 ± 2.84	0.515 ± 0.06	-0.217 ± 0.01
($\Delta X/X$ Low WV) %	34% ↓	11% ↓	143% ↑	56% ↓

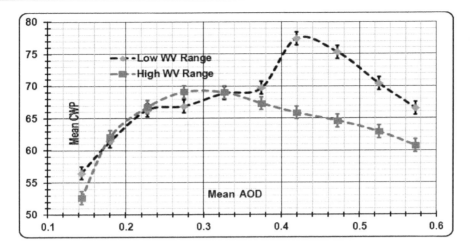

Figure 9. Response of cloud water path (CWP) to increasing aerosol optical depth (AOD) in low and high atmospheric WV environments for NAADR_Land.

4.1.3. AADR_Urban

For AADR in urban area (Figure 10), the threshold aerosol loading to shift from low to high water vapor environments is increased by approximately 3.2%. Meanwhile the corresponding CWP decreased significantly. In Table 4, CWP_{peak}, β_{drying} and $\beta_{moistening}$ decrease from low to high WV environments, respectively, by 20.2%, 82.2% and 36.1% of their initial magnitudes. Drying is the highest sensitivity process (highest β magnitudes) in both high and low water vapor environments.

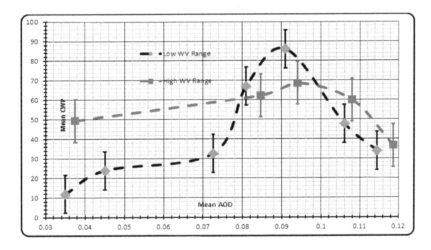

Figure 10. Response of cloud water path (CWP) to increasing aerosol optical depth (AOD) in low and high atmospheric water vapor (WV) environments for AADR_Urban.

Table 4. Comparative table of percent increase (↑) or decrease (↓) in the characteristics parameters of cloud water path response curve aerosol induced perturbation from low to high water vapor environment AADR_Urban.

Moisture	AOD_{peak}	CWP_{peak}	$\beta_{moistening}$	β_{drying}
Low WV	0.091 ± 0.031	86.0 ± 2.46	1.846 ± 0.180	-4.014 ± 0.180
High WV	0.094 ± 0.031	68.67 ± 2.46	0.328 ± 0.122	-2.563 ± 0.122
($\Delta X/X$ Low WV) %	3.2% ↑	20.2% ↓	82.2% ↓	36.1% ↓

4.1.4. AADR_SubTrop (Southern Africa)

In Figure 11, mean CWP increases with increasing mean AOD up to 0.095 and 0.084 (mean AOD values) for, respectively, low and high water vapor environments. In Table 5, Drying is the highest sensitivity mechanism in both low and high water vapor environment. β_{drying} magnitude is reduced by nearly 23% from low to high water vapor environment while $\beta_{moistening}$ is increased by 132%. The threshold aerosol loading necessary to shift from moistening to drying is decreased by approximately 13% from high to low water vapor environment.

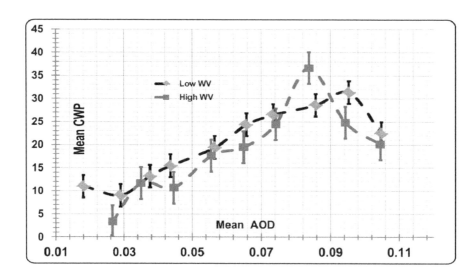

Figure 11. Response of cloud water path (CWP) to increasing aerosol optical depth (AOD) in low and high atmospheric water vapor (WV) environments for AADR_SubTrop.

Table 5. Percent increase (↑) or decrease (↓) in the characteristic parameters of cloud water path (CWP) response curve to increasing aerosol optical depth (AOD) for low and high atmospheric water vapor (WV) environments for AADR_SubTrop. The maximum and minimum changes are, respectively, observed for $\beta_{moistening}$ and AOD_{peak}.

Moisture	AOD_{peak}	CWP_{peak}	$\beta_{moistening}$	β_{drying}
Low WV	0.095 ± 0.028	31.35 ± 2.20	0.731 ± 0.238	-3.512 ± 0.238
High WV	0.084 ± 0.028	36.60 ± 2.20	1.693 ± 0.173	-2.719 ± 0.173
$(\Delta X/X \text{ Low WV}) \%$	13.1% ↓	16.7% ↑	131.6% ↑	22.6% ↓

4.1.5. AADR_Sahara

In Figure 12, CWP responses to AOD induced perturbation in high and low water vapor environments are nearly horizontal mirror images of each other. In Table 6, Drying and moistening are, respectively, the highest sensitivity process in low and high water vapor environments. From low to high WV environments, the maximum CWP corresponding to the lowest aerosol loading is reduced by approximately 30%.

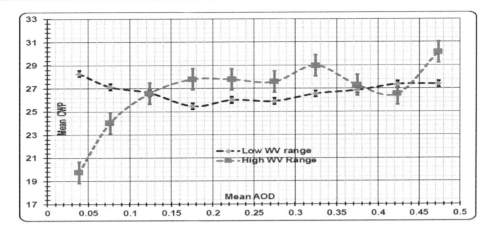

Figure 12. Response of cloud water path (CWP) to increasing aerosol optical depth (AOD) for low and high atmospheric WV environments for AADR_Sahara.

Table 6. Percent of decrease ($\downarrow$) in the characteristics parameters of cloud water path response curve to aerosol induced perturbation from low to high water vapor environment AADR_Sahara Drying is the highest sensitivity mechanism in low water vapor environment while moistening is the highest sensitivity mechanism in high water environment.

Moisture	AOD_{peak}	CWP_{peak}	$\beta_{moistening}$	β_{drying}
Low WV	0.04 ± 0.004	28.3 ± 2.20	--	-0.07 ± 0.238
High WV	0.04 ± 0.031	19.75 ± 2.20	0.230 ± 0.238	--
($\Delta X/X$ Low WV) %	$0\% \downarrow$	$30\% \downarrow$	--	--

4.2. Evaluation by Lower Tropospheric Static Stability (LTSS) Statistical Compositing

We study the sensitivity of CWP to AOD under conditions of varying lower tropospheric static stability (LTSS), comparing moistening and drying processes. LTSS data was collected from MODIS collection 5, where LTSS data is stored under stability indices, which are taken from ECMWF or NOAA NCEP reanalysis data. The total LTSS range for each study area is divided into two subranges (Low and High LTSS) of nearly equal counts based on LTSS histogram data distribution.

4.2.1. NAADR_Marine

In Figure 13, the response of CWP to AOD is non monotonic in both stable and unstable environments; the threshold aerosol loading necessary to shift from cloud moistening process to drying process is much larger in high stability environments than in low stability environments as well as the corresponding CWP. The magnitudes of cloud liquid water sensities δ during moistening in both stable and unstable environments are much larger than δ during drying The magnitude of $\beta_{moistening}$ is much larger in unstable environments than in stable environments. The magnitude β_{drying} is much larger in high stability environments than in low stability environments.

In Table 7, the transition from low to high static stability regime resulted in a significant increase in the parameters (AOD$_{peak}$, CWP$_{peak}$, β_{drying}) of cloud water path response curve to aerosol induced perturbation except for $\beta_{moistening}$ observed to decrease. β_{drying} increases nearly 273% from a low to high static stability regimes. The magnitude of $\beta_{moistening}$ is nearly 10 times β_{drying} in unstable atmospheric conditions.

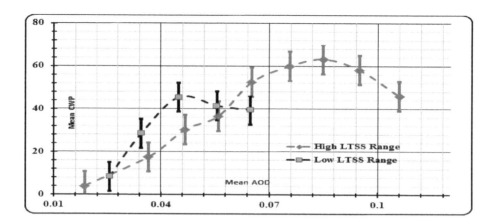

Figure 13. Response cloud water path (CWP) to increasing aerosol optical depth (AOD) in unstable and stable atmospheric environments for NAADR_Marine. The threshold AOD necessary to shift from moistening to drying processes is much larger in high LTSS environments than in low LTSS environments as well as the corresponding CWP.

Table 7. Percent of increase (↑) or decrease (↓) in the characteristics parameters of cloud water path response curve aerosol induced perturbation from low to high lower tropospheric static stability (LTSS) environments for NAADR_Marine. Except for $\beta_{moistening}$ every other parameter increased significantly from unstable to stable environments.

Stability	AOD_{peak}	CWP_{peak}	$\beta_{moistening}$	β_{drying}
Low LTSS	0.045 ± 0.011	45.42 ± 1.74	3.041 ± 0.27	-0.388 ± 0.27
High LTSS	0.085 ± 0.011	63.31 ± 1.74	1.958 ± 0.22	-1.446 ± 0.22
($\Delta X/X$ Low LTSS) %	88% ↑	39.4% ↑	35.6% ↓	273% ↑

4.2.2. NAADR_Land (Continental US)

In Table 8, except for the observed increase in AOD_{peak}, the transition from low to high static stability regime resulted in a significant decrease in the following keys parameters: CWP_{peak}, $\beta_{moistening}$, β_{drying}. The maximum decrease is observed in $\beta_{moistening}$ (89%) from low to high static stability regimes. The magnitude of $\beta_{moistening}$ in unstable atmospheric conditions is nearly five times the magnitude of β_{drying}. In stable atmospheric conditions, the magnitudes of both $\beta_{moistening}$ and β_{drying} are nearly identical. In Figure 14, for NAADR_Land, AOD_{peak} is much lower in low LTSS regimes than in high LTSS regimes.

Table 8. Percent of increase (↑) or decrease (↓) in the characteristics parameters of cloud water path response curve to aerosol induced perturbation from low to high lower tropospheric static stability (LTSS) environments for NAADR_Land. Except for AOD_{peak} observed to increase, every other parameter decreases from low to high LTSS.

Stability	AOD_{peak}	CWP_{peak}	$\beta_{moistening}$	β_{drying}
Low LTSS	0.243 ± 0.011	161 ± 1.74	11.15 ± 0.27	-2.67 ± 0.27
High LTSS	0.354 ± 0.011	70.0 ± 1.74	1.24 ± 0.22	-1.352 ± 0.22
($\Delta X/X$ Low LTSS) %	16.5% ↑	56.5% ↓	89% ↓	49.4% ↓

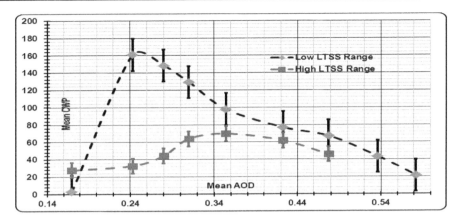

Figure 14. Response of cloud water path (CWP) to increasing aerosol optical depth (AOD) in unstable and stable atmospheric environments. For NAADR_Land, the threshold AOD necessary to shift from moistening to drying processes is much larger in stable environments.

4.2.3. AADR_Sahara

In Table 9, the magnitude of the moistening process is significantly higher than drying in high LTSS. The absence of data for low LTSS shows the prevalence of high static stability condition; this situation is consistent with the permanent high subsidence (high pressure) occurring in the Sahara region due the falling branches of Ferrell and Hadley cells. This lack of unstable conditions is reflected in Figure 15.

Table 9. Percent of increase (↑) or decrease (↓) in the characteristics parameters of cloud water path response to aerosol-induced perturbation for high lower tropospheric static stability (LTSS) environments for AADR_Sahara. No aerosol or cloud properties data were available for low static stability conditions.

Stability	AOD_{peak}	CWP_{peak}	$\beta_{moistening}$	β_{drying}
Low LTSS	--	--	--	--
High LTSS	0.221 ± 0.12	35.48 ± 1.95	0.591 ± 0.245	-0.026 ± 0.245
($\Delta X/ X$ Low LTSS) %	--	--	--	--

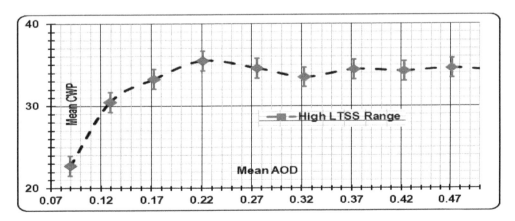

Figure 15. Response of cloud water path (CWP) to increasing aerosol optical depth (AOD) for high lower tropospheric static stability (LTSS) (stable atmospheric conditions). No data was available for low LTSS, possibly due high subsidence in the region.

4.2.4. AADR_SubTrop (off the coast of Southern African Agricultural Region).

In Figure 16, the threshold AOD necessary to shift from the moistening to drying is much larger in unstable atmospheric conditions than in stable atmospheric conditions. In Table 10, except for β_{drying} observed to increase in magnitude, the transition from low to high static stability regime resulted in a decrease in AOD_{peak}, CWP_{peak}, $\beta_{moistening}$. The increase in magnitude of β_{drying} and the decrease in magnitude of CWP_{peak} from a low to high static stability regime are very close.

Table 10. Percent of increase ($\uparrow$) or decrease ($\downarrow$) in the characteristics parameters of cloud water path response curve to aerosol induced perturbation from low to high lower tropospheric static stability (LTSS) environment for AADR_SubTrop. The relative sensitivities during drying in unstable and stable environment are very close while they were significantly different during moistening. All parameters are observed to decrease from an unstable to a stable environment except for β_{drying}.

Stability	AOD_{peak}	CWP_{peak}	$\beta_{moistening}$	β_{drying}
Low LTSS	0.085 ± 0.004	27.12 ± 0.51	0.186 ± 0.29	-0.311 ± 0.29
High LTSS	0.035 ± 0.004	24.76 ± 0.51	0.069 ± 0.29	-0.337 ± 0.29
($\Delta X/X$ Low LTSS) %	59% $\downarrow$	9.5% $\downarrow$	63% $\downarrow$	8.4% $\uparrow$

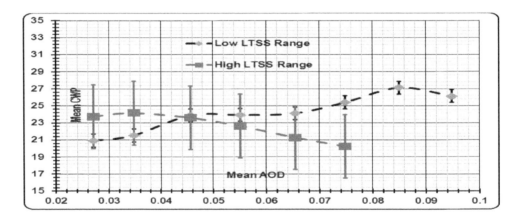

Figure 16. Response of cloud water path (CWP) to increasing aerosol optical depth (AOD) for low (unstable atmospheric condition) and high (stable atmospheric condition) lower tropospheric static stability (LTSS) for AADR_SubTrop.

4.2.5. AADR_URBAN

In Figure 17 for urban environments, moistening occurs only in unstable atmospheric condition. In stable atmospheric condition, drying is the highest sensitivity process. In Table 11, the threshold aerosol loading for AADR_Urban necessary to initiate the drying process is nearly 90% higher in low LTSS (unstable atmospheric conditions) than in high LTSS (stable atmospheric conditions). In low LTSS environment, the magnitude of relative sensitivity of CWP to aerosol induced perturbation during drying is significantly higher than its magnitude during moistening for AADR-Urban. In both Low and high LTSS environment, drying is the highest sensitivity process for AADR-Urban; no significant moistening process is occurring in high LTSS environment.

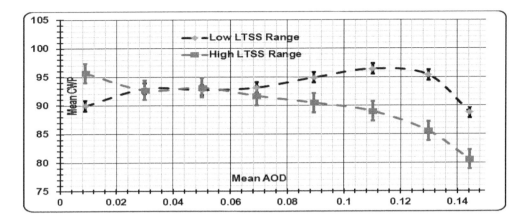

Figure 17. Response of cloud water path (CWP) to increasing aerosol optical depth (AOD) for low (unstable atmospheric condition) and high (stable atmospheric condition) lower tropospheric static stability (LTSS).

Table 11. Percent of increase (↑) or decrease (↓) in the characteristics parameters of cloud water path response curve aerosol induced perturbation from low to high lower tropospheric static stability (LTSS) environment for AADR_Urban.

Stability	AOD_{peak}	CWP_{peak}	$\beta_{moistening}$	β_{drying}
Low LTSS	0.11± 0.012	96.43 ± 1.95	0.024 ± 0.103	−0.285 ± 0.103
High LTSS	0.01 ± 0.012	95.68 ± 1.95	--	−0.047 ± 0.103
($\Delta X/X$ Low LTSS) %	91% ↓	0.7% ↓	--	83.5% ↑

4.3. Evaluation by Multilinear Regression Analysis (MLRA)

In this Section, CWP is simultaneously regressed on log (AOD), WV and LTSS. The objective of the MLRA is to untangle the influence of AOD, LTSS and WV on CWP during both moistening ($\beta > 0$) and drying ($\beta < 0$). This is done by quantifying the relative dependence of CWP on each of the three variables. Though the results generated by MLRA are known to be reliable, they still may be biased to a certain extent due to the sensitivity of MLRA to a number of factors, including:

(1) The nonlinear behavior of meteorological parameters in some regions (We suspect AADR_Sahara) may shift some meteorological influences to AOD [7].

(2) The influences of other variables unaccounted for in the context of the three variables may enhance or damp their individual influence.

(3) The direct relationship between AOD and WV observed in NAADR_Land, NAADR_Marine, AADR_SubTrop regions (Figure 18) may amplify the influence of AOD against the meteorological parameters or *vice versa*.

Though in these three regions Ridge Regression Analysis was used (a variant of the MLRA known to circumvent such situations), the possibility of error remains. Table 12 summarizes the relative influences of the three variables on CWP in the terms of regression coefficients on either side of the peak. The data is partitioned into low and high LTSS because Sections 4.1 and 4.2 showed it to have a greater effect than WV. Opposite signs in Table 12 may indicate possible antagonistic effects between variable(s) to produce $\beta > 0$ or $\beta < 0$ in either stable or unstable atmospheric conditions, while identical sign may indicates that the variable(s) synergistically worked together to produce $\beta > 0$ or $\beta < 0$.

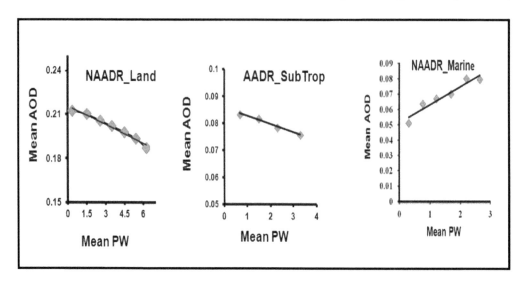

Figure 18. Linear relationship between aerosol optical depth (AOD) and Precipitable Water (PW) for NAADR_Marine, NAADR_Land and AADR_SubTrop. Ridge regression method is applicable for these three sites.

Table 12. Summary of the influences of all three variables on cloud water path (CWP) expressed as fraction the total influence (magnitudes of regression coefficients) during both moistening ($\beta > 0$) and drying ($\beta < 0$) and for each type of aerosol. Here it is assumed that the sum of the individual influence is equal to 1.

Region Type		Moistening ($\beta > 0$)			Drying ($\beta < 0$)		
		Log (AOD)	LTSS	WV	Log (AOD)	LTSS	WV
NAADR_Marine	Low LTSS	−0.52	0.39	0.09	0.33	0.50	0.17
	High LTSS	−0.52	−0.41	0.07	0.35	−0.33	−0.32
NAADR_Land	Low LTSS	0.04	0.93	−0.03	−0.05	0.92	0.03
	High LTSS	0.29	0.59	0.12	0.15	0.16	−0.69
AADR_Sahara	Low LTSS	0	−0.04	0.96	0.94	0.01	0.05
	High LTSS	0	0	−1	0.78	−0.18	−0.04
AADR_SubTrop	Low LTSS	−0.50	0.24	−0.26	0.69	0.26	0.05
	High LTSS	0.42	0.30	−0.28	0.36	−0.39	0.25
AADR_Urban	Low LTSS	0.39	0.11	0.5	0.21	0.36	0.43
	High LTSS	0.20	0.47	0.33	0.33	−0.6	−0.07

AADR_Urban:

Unstable Atmospheric Conditions (Low LTSS)

Both atmospheric AOD and WV concentrations are the governing factors that control the changes observed in CWP when $\beta > 0$. As aerosol loading increases, the strength of LTSS as controlling factor increases significantly while that of AOD and WV are observed to decrease but remain relevant. The three variables are responsible for producing $\beta < 0$.

Stable Atmospheric Conditions (High LTSS)

Atmospheric AOD , WV and LTSS are the governing factors that control the changes observed in CWP ($\beta > 0$). As aerosol loading increases ($\beta < 0$) in stable atmospheric conditions , the strength of both AOD and LTSS as controlling factor increase significantly while that of WV is observed to decrease and become irrelevant.

AADR_Sahara:

Unstable and Stable Atmospheric Conditions
Atmospheric WV and AOD are the governing factors that control changes induced in CWP.
(Respectively, $B > 0$ and $\beta < 0$). The effect of LTSS in producing either ($\beta > 0$) or ($\beta < 0$) is nearly zero, consistent with the fact that instability is not the prevailing condition in the region.

AADR_SubTrop:

Unstable and Stable Atmospheric Conditions
All three variables are governing factors in producing $\beta > 0$ or $\beta < 0$ except for WV in producing $\beta < 0$ in unstable atmospheres.

NAADR_Land:

In unstable environments, $\beta > 0$ and $\beta < 0$ processes are essentially governed by the LTSS. This is consistent with increased convective activity on land in summer. In addition, the contributions of both atmospheric WV content and AOD remain insignificant even after a substantial increase in aerosol loading. As the atmosphere becomes more and more stable, the atmospheric WV becomes the governing factors while aerosol loading as a factor gains in significance.

NAADR_Marine:

In both unstable and stable environment, $\beta > 0$ process is controlled by both LTSS and aerosol loading (AOD). The transition from an unstable environment to stable does not seem to significantly affect the impact of aerosol loading on changes that occur in CWP. In unstable and stable environment, $\beta < 0$ processes are essentially governed by all three parameters. However a significant contribution was recorded from LTSS while AOD and WV exhibited, respectively, moderate and low contributions.

5. Discussion

Observation based correlative studies in the past have exhibited contradictory results for cloud water path response to aerosol induced perturbation. In this study we have demonstrated that more attention should have been paid to how the correlations were actually established between cloud and aerosol proxies. Many past studies indirectly assumed a monotonic response of the cloud properties (CWP) to aerosol loading (AOD). The response of cloud water path to aerosol loading over the total AOD range shows an initial increase in CWP to a peak followed by a decrease over the total AOD range. This result is consistent with Xue et al. [40] who used Large Eddy Simulation to show an increase in CF (a proxy of CWP) to a maximum corresponding a threshold aerosol loading (100 cm^{-3}) followed by a decrease for aerosol loading greater than the threshold. Hoeve et al. [26] showed a peak at AOD ~0.3 in Modis COD (proxy for CWP) response to increasing AOD in Amazon biomass burning season.

Such response curves suggest that the same dataset could generate two or more different correlations between cloud variable (CWP) and aerosol loading proxies (AOD) depending on the AOD ranges the study is focused on. In that instance, there is a strong possibility that initially the divergences observed in some of the past correlative studies might have been simply an issue of analytical method [41,42].

One the advantages of this approach is that it offers the possibility to tune in meteorological parameters with cloud and aerosol proxies, determine the response curve for each meteorological setting and assess the variability in the shapes in terms of aerosol loading.

6. Conclusion

Modis aerosol and cloud observational data along with a proposed analytical method were used in this study to show the potential source of divergence in observation based past correlative studies. CWP response to increasing AOD is non monotonic and shows an initial increase to a maximum (peak) followed by a decrease. This feature persists over a range of meteorological conditions, and a variety of geographical locations and seasons. We advocate searching for a peak in the AOD-CWP curve, and confining any statistical analysis to data ranges on either side of this peak.

Acknowledgments

Data supporting this paper is Aqua Joint atmospheric product level 2 of MODIS collection 5. The data is available on MODIS website-NASA.

Author Contributions

Brian Vant-Hull had the original idea for the study, and Ousmane Sy Savane with all co-authors carried out the design. Ousmane Sy Savane was responsible for data cleaning and carried out the analyses. Ousmane Sy Savane drafted the manuscript, which was revised by all authors. All authors read and approved the final manuscript.

Conflicts of Interest

The authors declare no conflict of interest.

References

1. Lebsock, M.D.; Stephens, G.L.; Kummerow, C. Multisensor satellite observations of aerosol effects on warm clouds. *J. Geophys. Res.* **2008**, doi:10.1029/2008JD009876.
2. Nakajima, T.; Higurashi, A.; Kawamoto, K.; Penner, J.E. A possible correlation between satellite-derived cloud and aerosol microphysical parameters. *Geophys. Res. Lett.* **2001**, *28*, 1171–1174.
3. Bréon, F.M.; Tanre, D.; Generoso, S. Aerosol effect on cloud droplet size monitored from satellite. *Science* **2002**, *295*, 834–838.
4. Sekiguchi, M.; Nakajima, T.; Suzuki, K.; Kawamoto, K.; Higurashi, A.; Rosenfeld, D.; Sano, I.; Mukai, S. A study of the direct and indirect effects of aerosols using global satellite datasets of aerosol and cloud parameters. *J. Geophys. Res.* **2003**, doi:10.1029/2002JD003359.

5. Matsui, T.; Masunaga, H.; Kreidenweis, S.M.; Pielke, R.A., Sr.; Tao, W.-K.; Chin, M.; Kaufman, Y.J. Satellite-based assessment of marine low cloud variability associated with aerosol, atmospheric stability, and the diurnal cycle. *J. Geophys. Res.* **2006**, doi:10.1029/2005JD006097.

6. Han, Q.; Rossow, W.B.; Zeng, J.; Welch, R. Three different behaviors of liquid water path of water clouds in aerosol-cloud interactions. *J. Atmos. Sci.* **2002**, *59*, 726–735.

7. Kaufman, Y.J.; Koren, I.; Remer, L.A.; Rosenfeld, D.; Rudich, Y. The effect of smoke, dust and pollution aerosol on shallow cloud development over the Atlantic Ocean. *Proc. Natl. Acad. Sci.* **2005**, *102*, 11207–11212.

8. Albrecht, B. Aerosols, Cloud microphysics, and fractional cloudiness. *Science* **1989**, *245*, 1227–1230.

9. Ferek, R.; Garrett, T.; Hobbes, P.V.; Strader, S.; Johnson, D.; Taylor, J.; Nielson, K.; Ackerman, A.; Kogan, Y.; Liu, Q.; *et al.* Drizzle suppression in ship tracks. *J. Atmos. Sci.* **2000**, *57*, 2707–2728.

10. Vant-Hull, B.; Marshak, A.; Remer, L.; Li, Z. The effects of scattering angle and cumulus cloud geometry on satellite retrievals of cloud drop effective radius. *Geosci. Rem. Sens. Lett.* **2007**, *45*, 1039–1045.

11. Koren, I.; Kaufman, Y.J.; Remer, L.A.; Martins, V. Measurement of the effect of Amazon smoke on inhibition of cloud formation. *Science* **2004**, *303*, 1342–1345.

12. Taubman, B.A.; Marufu, L.; Vant-Hull, B.; Piety, C.; Doddridge, B.; Dickerson, R.; Li, Z. Smoke over haze: Aircraft observations of chemical and optical properties and the effects on heating rates and stability. *J. Geophys. Res.* **2004**, doi:10.1029/2003JD003898.

13. Ackerman, A.; Toon, O.B.; Stevens, D.E.; Heymsfield, A.J.; Ramanathan, V.; Welton, E.J. Reduction of tropical cloudiness by soot. *Science* **2000**, *288*, 1042–1047.

14. Koren, I.; Kaufman, Y.J.; Rosenfeld, D.; Remer, L.A.; Rudich, Y. Aerosol invigoration and restructuring of Atlantic convective clouds. *Geophys. Res. Lett.* **2005**, doi:10.1029/2005GL023187.

15. Burnet, F.; Brenguier, J.-L. Observational study of the entrainment-mixing process in warm convective clouds. *J. Atmos. Sci.* **2007**, *64*, 1995–2011.

16. Sinclair, V.A.; Gray, S.L.; Belcher, S.E. Controls on boundary layer ventilation: Boundary layer processes and large-scale dynamics. *J. Geophys. Res.* **2010**, doi:10.1029/2009JD012169.

17. Altaratz, O.; Bar-Or, R.Z.; Wollner, U.; Koren, I. Relative humidity and its effects on aerosol optical depth in the vicinity on convective clouds. *Environ. Res. Lett.* **2013**, *8*, 34025–34030.

18. Fan, J.; Leung, L.R.; Li, Z.; Morrison, H.; Qian, Y.; Zhou, Y.; Chen, H. Aerosol impacts on clouds and precipitation in southeast China–Results from bin and bulk microphysics for the 2008 AMF-China field campaign. *J. Geophys. Res.* **2012**, doi:10.1029/2011JD016537.

19. Lebo, Z.J.; Morrison, H.; Seinfeld, J.H. Are simulated aerosol induced effects on deep convective clouds strongly dependent on saturation adjustment? *Atmos. Chem. Phys. Discuss.* **2012**, *12*, 10059–10114.

20. Wang, H.; Feingold, G. Modeling mesoscale cellular structures and drizzle in marine stratocumulus. Part I: Impact of drizzle on the formation and evolution of open cells. *J. Atmos. Sci.* **2009**, *66*, 3237–3256.

21. Lebo, Z.J.; Feingold, G. On the relationship between responses in cloud water and precipitation to changes in aerosol. *Atmos. Chem. Phys.* **2014**, *14*, 11817–11831.

22. McComiskey, A.; Feingold, G. The scale problem in quantifying aerosol indirect effects. *Atmos. Chem. Phys.* **2012**, *12*, 1031–1049

23. Storelvmo, T.; Kristjánsson, J.E.; Myhre, G.; Johnsrud, M.; Stordal, F. Combined observational and modeling based study of the aerosol indirect effect. *Atmos. Chem. Phys.* **2006**, *6*, 3583–3601.

24. Storer, R.L.; Van den Heever, S.C.; Stephens, G.L. Modeling aerosol impacts on convective storms in different environments. *J. Atmos. Sci.* **2010**, *67*, 3904–3915.

25. Ackerman, A.; Kirkpatrick, M.P.; Stevens, D.E.; Toon, O.B. The impact of humidity above stratiform clouds on indirect aerosol climate forcing. *Nature* **2004**, *432*, 1014–1017.

26. Ten Hoeve, J.E.; Remer, L.A.; Jacobson, M.Z. Biomass burning aerosol effects on clouds. *Atmos. Chem. Phys.* **2011**, *11*, 3021–3036.

27. Han, Q.Y.; Rossow, W.B.; Chou, J.; Welch, R. Global variation of cloud effective droplet concentration of low-level clouds. *Geophys. Res. Letts.* **1998**, *25*, 1419–1422.

28. Han, Q.; Rossow, W.B.; Welch, R.M.; White, A.; Chou, J. Validation of satellite retrievals of cloud microphysics and liquid water path using observations from FIRE. *J. Atmos. Sci.* **1995**, *52*, 4183–4195.

29. Andreae, M.O. Correlation between cloud condensation nuclei concentration and aerosol optical thickness in remote and polluted regions. *Atmos. Chem. Phys.* **2009**, *9*, 543–556.

30. Squires, P. The microstructure and colloidal stability of warm clouds. *Tellus* **1958**, *10*, 262–271.

31. Rosenfeld, D.; Lensky, I.M. Satellite-based insight into precipitation formation in continental and maritime convective clouds. *Bull. Am. Meteorol. Soc.* **1998**, *79*, 2457–2476.

32. Wang, M.; Ghan, S.; Liu, X.; L'Ecuyer, T.S.; Zhang, K.; Morrison, H.; Ovchinnikov, M.; Easter, R.; Marchand, R.; Chand, D.; *et al.* Constraining cloud lifetime effects of aerosols using A-Train satellite measurements. *Geophys. Res. Lett.* **2012**, doi:10.1029/2012GL052204.

33. Terai, C.R.; Wood, R.; Leon, D.C.; Zuidema, P. Does precipitation susceptibility vary with increasing cloud thickness in marine stratocumulus? *Atmos. Chem. Phys.* **2012**, *12*, 4567–4583.

34. Mann, J.A.; Chiu, J.C.; Hogan, R.J.; O'Connor, E.J.; L'Ecuyer, T.S.; Stein, T.H.M.; Jefferson, A. Aerosol impacts on drizzle properties in warm clouds from ARM Mobile Facility maritime and continental deployments. *J. Geophys. Res.* **2014**, *119*, 4136–4148.

35. Duong, H.T.; Sorooshian, A.; Feingold, G. Investigating potential biases in observed and modeled metrics of aerosol-cloud precipitation interactions. *Atmos. Chem. Phys.* **2011**, *11*, 4027–4037.

36. Feingold, G.; McComiskey, A.; Rosenfeld, D.; Sorooshian, A. On the relationship between cloud contact time and precipitation susceptibility to aerosol. *J. Geophys. Res.* **2013**, *118*, 10544–10554.

37. Wen, G.; Marshak, A.; Cahalan, R.F.; Remer, L.A.; Kleidman, R.G. 3D aerosol-cloud radiative interaction observed in collocated MODIS and ASTER images of cumulus cloud fields. *J. Geophys. Res.* **2007**, doi:10.1029/2006JD008267.

38. Mauger, G.S.; Norris, J.R. Meteorological bias in satellite estimates of aerosol cloud relationships. *Geophys. Res. Lett.* **2007**, doi:10.1029/2007GL029952.

39. Eilers, P.H.C.; Goeman, J.J. Enhancing scaterplots with smoothed densities. *Bioinformatics* **2004**, *20*, 623–628.

40. Xue, H.; Feingold, G. Large-eddy simulation of trade wind cumuli: Investigation of aerosol indirect effects. *J. Atmos. Sci.* **2006**, *63*, 1605–1622.

41. Bony, S.; Dufresne, J.-L. Marine boundary layer clouds at the heart of tropical cloud feedback uncertainties in climate models. *Geophys. Res. Lett.* **2006**, *32*, L20806.

42. Tao, W.K.; Chen, J.P.; Li, Z.Q.; Wang, C.; Zhang , C.D. Impact of aerosols on convective clouds and precipitation. *Rev. Geophys.* **2012**, doi:10.1029/2011RG000369.

Permissions

The contributors of this book come from diverse backgrounds, making this book a truly international effort. This book will bring forth new frontiers with its revolutionizing research information and detailed analysis of the nascent developments around the world.

We would like to thank all the contributing authors for lending their expertise to make the book truly unique. They have played a crucial role in the development of this book. Without their invaluable contributions this book wouldn't have been possible. They have made vital efforts to compile up to date information on the varied aspects of this subject to make this book a valuable addition to the collection of many professionals and students.

This book was conceptualized with the vision of imparting up-to-date information and advanced data in this field. To ensure the same, a matchless editorial board was set up. Every individual on the board went through rigorous rounds of assessment to prove their worth. After which they invested a large part of their time researching and compiling the most relevant data for our readers.

The editorial board has been involved in producing this book since its inception. They have spent rigorous hours researching and exploring the diverse topics which have resulted in the successful publishing of this book. They have passed on their knowledge of decades through this book. To expedite this challenging task, the publisher supported the team at every step. A small team of assistant editors was also appointed to further simplify the editing procedure and attain best results for the readers.

Apart from the editorial board, the designing team has also invested a significant amount of their time in understanding the subject and creating the most relevant covers. They scrutinized every image to scout for the most suitable representation of the subject and create an appropriate cover for the book.

The publishing team has been an ardent support to the editorial, designing and production team. Their endless efforts to recruit the best for this project, has resulted in the accomplishment of this book. They are a veteran in the field of academics and their pool of knowledge is as vast as their experience in printing. Their expertise and guidance has proved useful at every step. Their uncompromising quality standards have made this book an exceptional effort. Their encouragement from time to time has been an inspiration for everyone.

The publisher and the editorial board hope that this book will prove to be a valuable piece of knowledge for researchers, students, practitioners and scholars across the globe.

List of Contributors

Fong Ngan and Xinrong Ren
Air Resources Laboratory, National Oceanic and Atmospheric Administration, 5830 University Research Court, College Park, MD 20740, USA
Cooperative Institute for Climate and Satellites, University of Maryland,
5825 University Research Court, College Park, MD 20740, USA

Mark Cohen, Winston Luke and Roland Draxler
Air Resources Laboratory, National Oceanic and Atmospheric Administration, 5830 University Research Court, College Park, MD 20740, USA

Kunshan Bao and Ji Shen
State Key Laboratory of Lake Science and Environment, Nanjing Institute of Geography and Limnology, Chinese Academy of Sciences, Nanjing 210008, China

Guoping Wang
Key Laboratory of Wetland Ecology and Environment, Northeast Institute of Geography and Agroecology, Chinese Academy of Sciences, Changchun 130102, China

Gaël Le Roux
Université de Toulouse, INP, UPS, EcoLab (Laboratoire Ecologie Fonctionnelle et Environnement), ENSAT, Avenue de l'Agrobiopole, 31326 Castanet Tolosan, France
CNRS, Eco Lab, 31326 Castanet Tolosan, France

Paolo Martano
CNR−Istituto di Scienze dell'Atmosfera e del Clima, U.O. S. Lecce, Via Monteroni, 73100 Lecce, Italy

Stefano Falcinelli
Department of Chemistry, Stanford University, Stanford, CA 94305, USA
Department of Civil and Environmental Engineering, University of Perugia, Via G. Duranti 93, Perugia 06125, Italy

Fernando Pirani
Department of Chemistry, Biology and Biotechnologies, University of Perugia, Via Elce di Sotto 8, Perugia 06123

Franco Vecchiocattivi
Department of Civil and Environmental Engineering, University of Perugia, Via G. Duranti 93, Perugia 06125, Italy

Hongmei Xu
Department of Environmental Science and Engineering, Xi'an Jiaotong University, Xi'an 710049, China
Key Lab of Aerosol Chemistry & Physics, Institute of Earth Environment, Chinese Academy of Sciences, Xi'an 710061, China

Benjamin Guinot
Laboratoire d'Aerologie, Observatory Midi-Pyrenees, CNRS−University of Toulouse, Toulouse 31400, France

Zhenxing Shen
Department of Environmental Science and Engineering, Xi'an Jiaotong University, Xi'an 710049, China

Kin Fai Ho
School of Public Health and Primary Care, The Chinese University of Hong Kong, Hong Kong, China

Xinyi Niu
Key Lab of Aerosol Chemistry & Physics, Institute of Earth Environment, Chinese Academy of Sciences, Xi'an 710061, China
School of Human Settlements and Civil Engineering, Xi'an Jiaotong University, Xi'an 710049, China

Shun Xiao
Key Lab of Aerosol Chemistry & Physics, Institute of Earth Environment, Chinese Academy of Sciences, Xi'an 710061, China
Shaanxi Meteorological Bureau, Xi'an 710014, China

Ru-Jin Huang
Key Lab of Aerosol Chemistry & Physics, Institute of Earth Environment, Chinese Academy of Sciences, Xi'an 710061, China
Laboratory of Atmospheric Chemistry, Paul Scherrer Institute (PSI), 5232 Villigen, Switzerland
Centre for Climate and Air Pollution Studies, Ryan Institute, National University of Ireland Galway, Galway, Ireland

Junji Cao
Key Lab of Aerosol Chemistry & Physics, Institute of Earth Environment, Chinese Academy of Sciences, Xi'an 710061, China
Institute of Global Environmental Change, Xi'an Jiaotong University, Xi'an 710049, China

Hanbin Zhang
College of Atmospheric Science, Nanjing University of Information & Science Technology, Nanjing 210044, China; E-Mail: zhb828828@163.com

Jing Chen
Center of Numerical Weather Prediction of CMA, Beijing 100081, China

Xiefei Zhi
Key Laboratory of Meteorological Disaster, Ministry of Education College of Atmospheric Science, Nanjing University of Information & Science Technology, Nanjing 210044, China

Yanan Wang
Center of Meteorological Service of Zhejiang, Hangzhou 310017, China

Giovanni Pitari, Glauco Di Genova and Natalia De Luca
Department of Physical and Chemical Sciences, Università degli Studi dell'Aquila, Via Vetoio, Coppito, 67100 L'Aquila, Italy

Shuhui Zhao
Key Lab of Global Change and Marine-Atmospheric Chemistry of State Oceanic Administration, Third Institute of Oceanography, State Oceanic Administration, Xiamen 361005, China
Institute of Urban Environment, Chinese Academy of Sciences, Xiamen 361021, China
College of Ocean and Earth Sciences and State Key Laboratory of Marine Environmental Science, Xiamen University, Xiamen 361005, China

Liqi Chen
Key Lab of Global Change and Marine-Atmospheric Chemistry of State Oceanic Administration, Third Institute of Oceanography, State Oceanic Administration, Xiamen 361005, China
College of Ocean and Earth Sciences and State Key Laboratory of Marine Environmental Science, Xiamen University, Xiamen 361005, China

Yanli Li and Zhenyu Xing
Institute of Urban Environment, Chinese Academy of Sciences, Xiamen 361021, China

Ke Du
Department of Mechanical and Manufacturing Engineering, University of Calgary, Calgary, AB T2N 1N4, Canada

Ousmane Sy Savane
Earth and Environmental Sciences Department, Graduate Center, The City University of New York, New York, NY 10031, USA

Brian Vant-Hull, Shayesteh Mahani and Reza Khanbilvardi
Cooperative Remote Sensing Science and Technology (CREST) Institute, City University of New York, New York, NY 10031, USA